AIRCRAFT WELDING

Prepared by

NAVAL AIR TECHNICAL TRAINING COMMAND

For Publication by the

BUREAU OF NAVAL PERSONNEL

NAVY TRAINING COURSES
NAVPERS 10322-A

UNITED STATES
GOVERNMENT PRINTING OFFICE
WASHINGTON : 1953

PREFACE

There is an almost limitless variety of welding jobs on such diverse aircraft parts as bearings, tanks, gasoline and oil lines, generator housings, gun mounts, landing gear units, and engine valves.

This book, written primarily for the welding phase of the Aviation Structural Mechanic rating, deals with both oxy-acetylene and electric arc welding. The section devoted to the oxyacetylene method begins with an introduction to fundamental equipment, and is followed by practical instructions in welding techniques. Ensuing chapters deal with oxyacetylene cutting, brazing, soldering, and hand forging— all subjects of related importance to the Aviation Structural Mechanic.

The second section deals with arc welding, inert arc welding, and atomic-hydrogen welding.

The discussions on arc welding begin with equipment and its operation, then move on to studies of the arc and electrodes. Information on factors governing arc welded joints and techniques of position welding follow. This section is concluded with a discussion of expansion, contraction, and distortion, and the arc welding of nonf errous metals.

The chapter on inert arc welding briefly defines the process and discusses types, uses, and care of equipment and material. Simple jobs, including current setting, gas adjustment, and techniques involved in developing skills are also included.

The chapter on atomic hydrogen welding discusses theory of operation, types of equipment, and various applications that are possible with this method of welding. A few practice examples are included to familiarize the student with welding proccesses and techniques.

Grateful acknowledgment is made for the technical information supplied by the Lincoln Electric Company for arc welding, the Linde Air Products Company and the Miller Electric Manufacturing Company for inert arc welding, and the General Electric Company for atomic-hydrogen welding data.

As one of the navy training courses, this book represents the joint endeavor of the Naval Air Technical Training Command and the Training Publications Section of the Bureau of Naval Personnel.

STUDY GUIDE FOR AVIATION STRUCTURAL
MECHANIC

The table below indicates which chapters of this book apply to your rate. To use the table, follow these rules:

1. Select the column which applies to your rating. If you are in the Regular Navy you will use the column headed "AM," which is the general sebvice batino. If you are an AMS in the Naval Reserve, you will use the column headed by your particular emergency service rating. Aviation Structural Mechanics H (Hydraulic Mechanics) are not required to know welding to advance in rating.

2. Observe chapter numbers appearing opposite the rate to which you are seeking advancement

3. Study those particular chapters. They include information which will assist you in meeting the qualifications for your rating. (See appendix III of this book for a complete list of qualifications for advancement in rating.) To gain a well-rounded view of the duties of a welder,

it is recommended that you read the other chapters of this book even though they do not pertain directly to your rate.

READING LIST

NAVY TRAINING COURSES

Aircraft Structures, NavPers 10331-A Aircraft Materials, NavPers 10330 Aircraft Hydraulics, NavPers 10332-A Introduction to Aircraft, NavPers 10308-A Mathematics, Vol. 1, NavPers 10069-A Mathematics, Vol. 2, NavPers 10070-A Blueprint Reading, NavPers 10077 Hand Tools, NavPers 10308-A

United States Armed Forces Institute (usafi) courses for additional reading and study are available through your Educational Services Officer.* A partial list of those courses applicable to your rate follows:

•"Members of the United States Armed Forces Reserve components, when on active duty, are eligible to enroll for usafi courses, services, and materials if the orders calling them to active duty specify a period of 120 days or more, or if they have been on active duty for a period of 120 days or more, regardless of the time specified In the active duty orders."

USAFI TEXTS

Number X 204 X 205 J 362 J 363

Aircraft Maintenance and Repair I Aircraft Maintenance and Repair II Arc Welding Gas Welding

Title

CONTENTS

AIRCRAFT WELDING

CHAPTER 1

FUNDAMENTALS OF WELDING

AIRCRAFT MAINTENANCE AND WELDING

The Aviation Structural Mechanic is one of the most important members of the Navy's aircraft maintenance crews. His duties consist of exactly what the title of his rating implies—the maintenance and repair of aircraft structural elements.

A knowledge of aircraft welding is of primary importance to the Aviation Structural Mechanic, since this art is an indispensable part of aircraft maintenance work. In the performance of his duties, welding—and that means all phases of the crafts-constitutes one of the main elements of his specialty.

The repair of welded aircraft structures may be divided into two general classifications— the repair of structural welded parts and assemblies, and the repair of nonstructural parts and assemblies. The former category includes welded steel fuselages, landing gears, tail surfaces, and engine mounts, while nonstructural parts refer to aluminum alloy cowlings, fuel tanks, and exhaust stacks and manifolds which, are fabricated from steel of the stainless or heat-resistant types.

Welding is used not only for the repairs mentioned in the preceding paragraph, but also for fabricating jigs and fixtures for the repair shop and for repairing heavy machinery. In the aviation structural mechanic schools, instruction predominates in-the oxyacetylene method, as this process is employed in the welding of the majority of aircraft metals. A brief comparison of the different welding processes will provide a background for the intensive study of oxyacetylene welding.

WELDING DEFINED

In general, welding consists of the controlled melting or fusing of adjacent edges or surfaces of metals so that the molten portions flow together into a common mass of metal. Upon cooling, the fused portions are united into a castlike structure which binds or joins the surrounding areas of metal.

Welding may be divided into two general classifications— pressure and nonpressure. In pressure welding, the surfaces to be joined are heated until they are in a plastic or semimolten stage, and are then joined by applied forces. Examples of this class of welding are forge, spot, and resistance welding. In nonpressure welding, the surfaces to be fused are brought to a molten state thus causing them to flow together. In this latter class may be found thermit, ELECTRIC ARC, OXYHYDROGEN, and OXYACETYLENE Welding.

WELDING PROCESSES DEFINED

The following brief explanations of the different welding processes are taken from the tenth edition of the Welding Encyclopedia.

Forge welding is a process of welding metals in the plastic state by means of manual or mechanical hammering. This process includes blacksmith welding, hammer welding, and roll welding.

Spot welding is a resistance welding process wherein the weld is made in one or more spots by the localization of the electric current between contact points.

Resistance welding is a pressure welding process wherein the welding heat is obtained by passing an electric current across the resistance set up between the contact area to be welded.

Thermit welding is the process of heating aluminum and iron oxide to approximately 5,000° F. These molten oxides are applied to the metal to be welded at which time a chemical reaction fuses the filler metal to the preheated base metal.

Arc welding is the concentration of the heat in an electric circuit at an air or gas gap for the purpose of raising metals to be joined to a melting temperature.

Oxyhydrogen welding is accomplished by using a torch flame to heat the metal. The flame is produced by the combustion of a mixture of one volume of oxygen with four volumes of hydrogen in order to prevent oxidation of the metal. The temperature of the flame is about 4,300° F.

Oxyacetylene welding is a process of f using metals together with heat produced by the combustion of oxygen and acetylene. The chemical union of these two gases produce the hottest flame known to man—about 6,300° F. The oxyacetylene process is widely used in aircraft welding because of this high temperature, and because the flame is easy to regulate. Oxygen and acetylene are easy to produce in large quantities and can be transported safely when properly confined.

PROPERTIES OF METALS

Because of the relatively low safety factor in airplane structures, all metals used in aircraft construction will possess greater strength in proportion to weight. They will also possess other properties in widely different degrees because of the varying uses to which they are put. A general knowledge of the properties of metals—especially as they are affected by welding—will enable the Aviation Structural Mechanic to make a repair acceptable in all respects because most of the original qualities of the metal are left intact

It is suggested that this discussion be studied in conjunction with the Navy Training Course, Aircraft Materials (NavPers 10330) in order that a more thorough understanding of aircraft metal properties may be gained.

Some of the properties of metals, by which their adaptability to aircraft may be evaluated, are:

1. Strength. 5. Fatigue resistance.
2. Elasticity. 6. Durability.
3. Ductility. 7. Hardness.
4. Toughness. 8. Ease of fabrication.

STRENGTH

Aircraft metals must have a high strength-to-weight ratio. The aircraft structure, exclusive of engine, tanks, etc., represents from 25 to 35 percent of the total weight of the craft, despite every refinement in design. It is obvious that the weight of every separate part must be kept to a minimum.

The advantage of one aircraft material over another increases approximately as the square of the strength-to- weight ratio. It is this characteristic that makes aluminum so valuable, since it is only one-third as heavy as steel. High-strength steel alloys also possess a high strength-to-weight ratio, but they are rather difficult to form into the thin cross sections which must be used to attain a low weight comparable to that of aluminum.

FATIGUE RESISTANCE

Another factor which is particularly important in the metals used in aircraft construction is fatigue resistance. Aircraft structures must withstand repeatedly applied loads or reversals of loading, and are continually subjected to severe vibration. To avoid failure of planes in service, welds on structural parts must be made in such a way as to maintain the fatigue resistance of the parts.

RELATION OF QUALITIES

The qualities of elasticity, ductility, toughness, and hardness are closely related, and an increase in one quality may often be gained only at the expense of a reduction in one or more of the other properties. For example, hardened carbon steel is less elastic and ductile than the same steel in a tempered condition. Stainless steel, however, is likely to have most of these qualities to a high degree.

DURABILITY

For military aircraft, the adaptability of metals to service conditions is almost equal in importance to their physical characteristics. It is desirable that the material used in an airplane be affected as little as possible by changes in climatic conditions, or unfavorable storage facilities. In Navy planes, resistance to the corrosive action of salt spray is a highly important quality.

Examples of a metal selected for its durability are Alclad sheet and stainless steel. Alclad sheet is a light metal used for airplane skin because of its corrosion resisting qualities, and stainless steel is used in exhaust collectors because of its heat-resisting properties.

Since heat affects the qualities of resistance to corrosion of all aircraft metals, welding must be performed in such manner as to avoid seriously impairing these qualities. It is also necessary that these qualities be restored in some cases by the proper heat treatment when the welding is completed.

EASE OF FABRICATION

Aircraft materials must obviously be easy to bend or to form. For this reason, aluminum and its alloys are used for cowl rings; for skin; on fuselage and wings; and also for formed or extruded ribs, spars, or bulkheads. Chrome molybdenum steel is also quite readily formed into desired shapes and is used for structural members where great strength is required. Metals for use in aircraft must also be easy to rivet or weld, since speed and ease of assembly are important factors in their usefulness.

AIRCRAFT METALS

■

Some of the metals used in aircraft which require welding in fabrication or repair are carbon steel, chrome molybdenum steels, and stainless steels. There are other metals which are more infrequently used and which do not require welding in their repair, that can be listed as aircraft metals. These are brass, copper, and bronze.

CARBON STEEL

Although aluminum, with its alloys, is the metal used to the greatest extent in aircraft construction, steel and its alloys still comprise about 20 percent of the structural weight of the airplane. Carbon steels were at one time the only metals of this type used for structural parts. At present, these steels have been largely superseded by chrome molybdenum steel.

All steels contain carbon in varying amounts, but only those containing no other alloying element are referred to as carbon steel. The steels which are generally used for parts fabricated by welding are those containing not more than 0.30 percent carbon. Due to the fact that molten steel has a great affinity for carbon, extreme care must be exercised during welding to avoid change in its content.

Those steels containing less than 0.30 percent carbon are known as low carbon steels. They are comparatively elas-

tic, tough, and ductile, but are not exceptionally hard or strong. They are easily welded by all processes, and the resulting welds or joints are of extremely high quality.

Medium carbon steels— containing from 0.30 to 0.50 percent carbon—are hardened as a result of the increased amount of carbon present. They can be welded fairly easily by the oxyacetylene process. In some cases, preheating may be necessary, in addition to heat treatment after welding, to produce the desired weld quality. This is especially true for steels containing over 0.40 percent carbon.

High carbon steels, containing from 0.50 to 0.90 percent carbon, are harder and more brittle because of the higher carbon content. These steels are generally heat treated because they are used for tools where hardness and strength are the desired qualities. The welding will affect the heat treatment and produce a joint of different properties than those possessed by the original metal. Care should be taken to prevent overheating of these parts, and the weld should be completed as quickly as possible. Materials to be welded are often preheated to speed up the welding process.

CHROME MOLYBDENUM STEEL

Chrome molybdenum steel has almost completely replaced carbon steel in such applications as fuselage tubing, landing gear struts, and other structural parts because of its high strength, ease of forming, and adaptability to welding.

A much-used series of chrome molybdenum steel has a carbon range of 0.25 to 0.55 percent, molybdenum 0.15 to 0.25 percent, and chromium 0.50 to 1.10 percent. These steels, when suitably heat treated, are deep hardening, easily machined, and readily welded by either gas or electric methods.

Because chrome molybdenum steel is thin walled, it air-hardens readily, resulting in high tensile strength and low ductility. In welding it, consideration must be given to expansion and contraction, and to avoiding undue stress on the hot metal.

288500°—63 2

STAINLESS STEELS—CHROME NICKEL ALLOYS

Alloys in this classification contain sufficient nickel and manganese, in addition to the chromium (with or without small amounts of other metals), to maintain an essentially austenitic condition. In other words, stainless steels have such a grain structure at ordinary temperatures that they possess low heat conductivity, comparatively high coefficient of expansion, nonmagnetic properties, and satisfactory toughness.

All stainless steels, except those stabilized by colombium or titanium, when exposed to temperatures of 700° to 1,500° F., will show an accumulation of carbides at the grain boundaries. This is known as carbide precipitation and is found in stainless steel that has been held in this temperature range. It is found also in lines approximately parallel to the weld bead up to a quarter-inch distance even in stainless steel plates that have been properly heat-treated after welding. This is the result of the heat of the welding operation, and is evidence that the plate metal in that narrow zone has been heated to a temperature range wherein carbide precipitation can take place.

Stainless steels containing colombium, titanium, or molybdenum are "stabilized" (to dissolve precipitated carbides and to prevent intergranular corrosion) by heating to 1,550°-1,625° F., for from 2 to 4 hours, and then quenching, as in annealing. This is done after the welding operation has been completed.

Stainless steel will conduct heat only about 40 percent as fast as mild steel, but its coefficient of expansion is about 50 percent greater than mild steel. Since the distribution of heat from the acetylene flame is wider than that obtained from arc methods, it is desirable to use chill blocks to absorb the heat when welding with gas. Since most of the stainless steel used in aircraft applications is one-sixteenth inch or less, extensive use of jigs and clamps will be necessary to control buckling and warping.

EXPANSION AND CONTRACTION

The principles of expansion and contraction of metals is based upon facts which are easily understood. When metal is heated, it expands appreciably in all its dimensions. When metal cools, it becomes reduced in size—it contracts.

Steel rails in a railroad track show clearly the effect of changes in temperature upon metal. Summer heat expands the rails until the ends of all sections of the rails are in contact ; low winter temperatures cause the rails to contract until there is a sizeable gap between the ends.

The extreme range in temperature to which welded parts are subjected creates a serious problem for the welder. The forces of expansion and contraction cannot be eliminated or mechanically controlled. If the welder is to produce satisfactory welds, he must have a knowledge of the amount of expansion and contraction encountered in the different metals he must weld. He must also have a knowledge of the effects these forces have on different thicknesses and compositions of metals, and of the steps which can be taken to counteract or compensate for these forces.

The rate of expansion of different metals is constant and the amount of expansion varies directly with the temperature. Scientists call the rate of expansion "coefficient of expansion." This term refers to the amount that a unit length of metal will increase in length if the temperature is raised 1° F. The amount that a given metal will expand can be readily calculated from the figures given in table 1. For example, if a 10-foot low carbon rod is heated from 70° to 2,700°, the increase in length is equal to $0.00000630 \times 2630 \times 120$, which equals 1.99 inches.

Not only is the amount of expansion different for each metal, but the rate at which the heat is conducted is also quite different for each of the metals which must be welded. There are other factors which influence the amount of expansion or contraction in a welded structure or part, such as thickness of the part, the speed of welding, and the method

Table 1.—EXPANSION OF METALS

of clamping. If the parts being welded are free to expand in all directions, the welder will usually have little trouble, as the parts will expand equally, and in cooling can return to their original dimensions. In general, more trouble is due to contraction while the metal is cooling. The weld

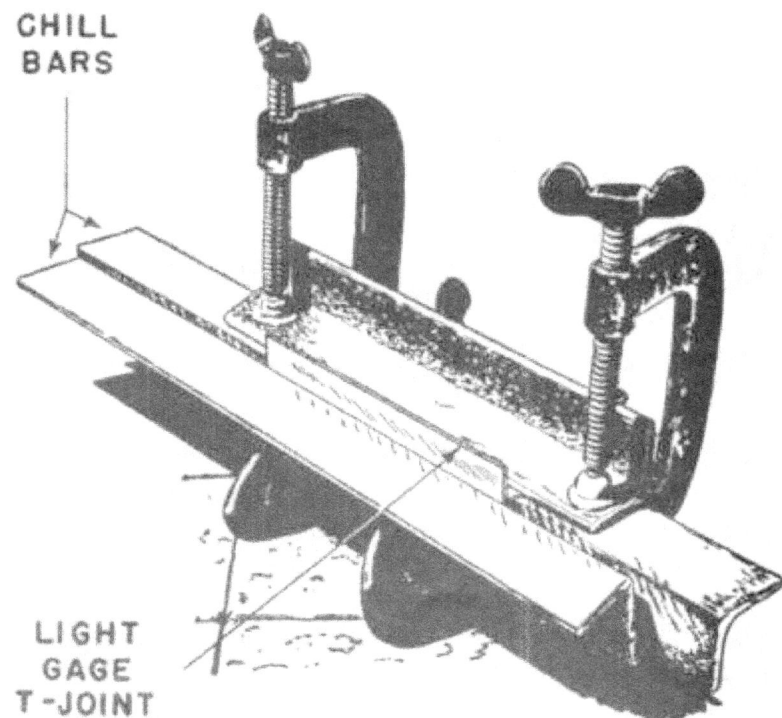

CHILL BARS

LIGHT GAGE T-JOINT

Figure 1.—Angle iron used as both heal conductor and jig for a light-gage T-joint.

metal is so soft that expansion is taken care of by a thickening or upsetting of the metal. If the welder understands what effects expansion and contraction will have upon a particular job, he should generally be able to devise some means of compensating for these forces.

The effect of expansion and contraction due to the heat from the welding flame on thin sheet (one-eighth inch or less in thickness) is a tendency to develop buckling or warping. This buckling or warping is a result of the greater amount of surface area in proportion to the weight. The welding heat spreads rapidly, and the metal cools very quickly when the source of heat is removed.

Expansion on thin sheets may be controlled somewhat by tacking along the joint, but the most effective means of alleviating the warping is to remove some of the heat from the metal near the weld. Heavy pieces of metal, known as chill bars, may be placed on either side of the seam, as shown in figure L Welding jigs utilize this same principle of removing the heat from the base metal.

Long seams (over 10 to 12 inches) have a tendency to draw together as the weld progresses. The spot being welded is melted so rapidly that most of the expansion is taken care

EFFECT ON SHEET METAL

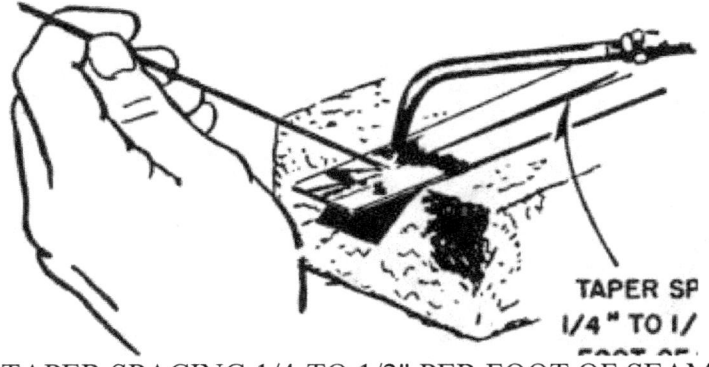

TAPER SPACING 1/4-TO 1/2" PER FOOT OF SEAM.

Figure 2.—Allowance for contraction.

of in the molten metal. As the metal cools, it contracts and tends to draw the seam together. This can be overcome by setting up the parts to be welded with an increased allowance at the far end, which varies according to the metal used, and its thickness. This allowance per foot of seam is generally one-fourth to one-half inch per foot of seam for steel. This principle is illustrated in figure 2.

Sheet metal under one-sixteenth inch in thickness is best handled by flanging the edges, tack-welding at intervals along the seam, and then welding.

PLATES

There is less tendency to warp and buckle when plate (over %-inch stock) is welded, because of the greater proportion of metal to surface area, which limits the flow of heat from the welded area. However, it is advisable to allow the same amount for contraction along a long straight seam as was allowed on thin sheet. Preheating will help to prevent cracks caused by uneven expansion and contraction.

TUBULAR STRUCTURES

A welded tubular structure will have terrific stresses set up by the welding, if preventive measures are not taken. When welding two members of a T-joint, as shown in figure 3, tube (A) tends to draw up. This is due to the fact that most of the expansion is taken care of in the molten metal, but as the weld cools, the base metal contracts beyond its original shape.

Buckling can be relieved by preheating before the welding

Figure 3.—Effect of Improper welding technique on tubular structures.

operation begins. Contraction still takes place at the weld, but there is also shrinking in the rest of the structure at approximately the same rate. Internal stress at the weld is thereby relieved.

QUIZ

1. What is welding?
2. Name seven welding processes.
3. What types of steels are used in aircraft construction ?
4. What are the chief characteristics of low carbon steel ?
5. Name the chief characteristics of chrome molybdenum

steel.

6. Name the chief characteristics of stainless steels.

7. What does "coefficient of expansion" mean ?

8. How may heat be taken away from a sheet metal joint weld ?

9. What should be done before welding heavy plate?

10. What should be done before welding tubular structures ?

CHAPTER 2

INTRODUCTION TO OXYACETYLENE WELDING EQUIPMENT

The amount of equipment required for successful oxy-acetylene welding operations is considerable, and may be either portable or stationary. The portable equipment, illustrated in figure 4, is generally fastened on a hand truck so that it may be moved from place to place as needed. It consists of the following elements: two cylinders, one containing oxygen and the other acetylene; a welding torch with necessary tips and mixing head; lengths of green or black hose for oxygen, and red or maroon hose for acetylene;

OXVOCN PRESSURE HEGULATOH

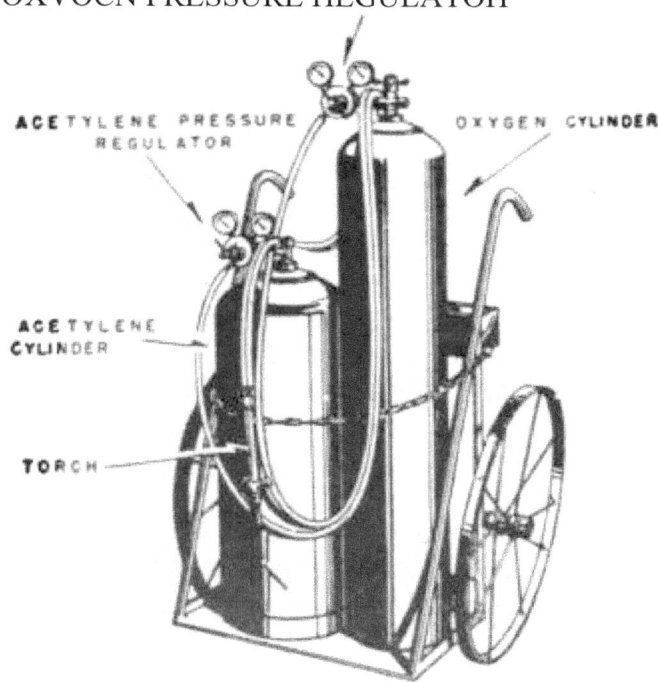

Figure 4.—Portable oxyacetylene welding equipment.

and miscellaneous equipment such as goggles, friction LIGHTER, WRENCH, and FIRE EXTINGUISHER.

If the shop has stationary equipment, it will resemble that shown in figure 5, where the oxygen and acetylene are piped to a number of welding stations. In stationary units, the oxygen and acetylene cylinders are connected to a manifold. Each bank of cylinders has a master regulator which governs the flow of the gases. The manifold is the long pipe in figure 5 directly above the cylinders into which the short pipe from each cylinder feeds.

Each system is equipped with a hydraulic back pressure valve located between the manifold and station line. This device is a water seal valve which prevents the gases—or a flashback—from going back through the system to the cylinders. In some cases, the acetylene is piped directly from an acetylene generator rather than from a cylinder.

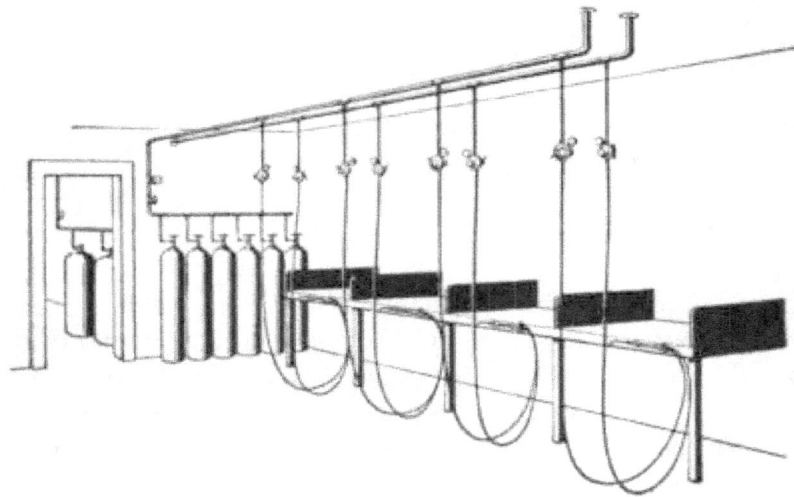

Figure 5.—Stationary oxyacetylene welding equipment.

Oxygen, an odorless, tasteless, colorless gas, is needed in welding to mix with the acetylene and cause the gas to burn at a temperature high enough to melt the metal being welded. Oxygen itself does not burn—rr supports combustion. The oxygen in the air is that element causing any inflammable substance to burn.

Air is about one-fifth oxygen, the rest being nitrogen with a small percentage of rarer gases such as argon, neon, and helium. These gases in the-air are simply a scrambled mixture, with no chemical union, although special machinery and processes are required to separate them and to transform air from a gas into a liquid.

When this separation and transformation occurs, the liquid air obtained is an extremely cold mixture of liquid oxygen and liquid nitrogen. Oxygen is isolated from the nitrogen by a process known as rectification, which delivers high purity oxygen to a storage holder from which it is compressed into steel cylinders for shipment.

17

OXYGEN CYLINDERS

Typical oxygen cylindebs, such as shown in figure 6, are made of seamless drawn steel. The Navy standard oxygen cylinder for welding and cutting operations is approximately 9 inches in diameter and 51 inches long, holding 200 cubic feet of oxygen at 1,800 pounds of pressure.

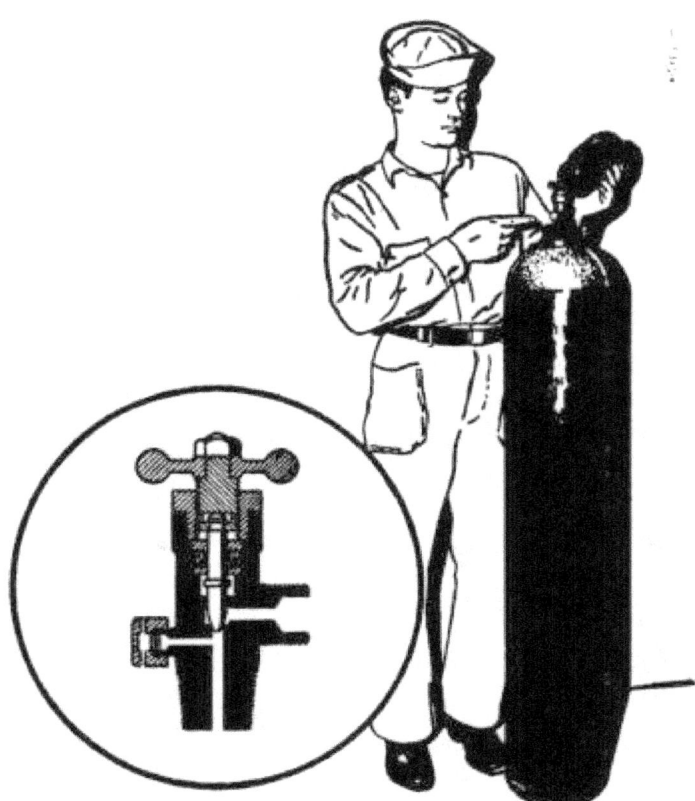

Figure 6.—Th« oxygen cylinder valve with creu-Mctioiial view (hewn in insnt.

Oxygen cylinders are painted with identifying colors— green for technical oxygen and green with a white band at the top for aviators' breathing oxygen. Both are used in shipboard welding, but the technical oxygen is more commonly used ashore. The top of each cylinder is equipped with a

h igh -pressure valve which controls the flow of oxygen. This valve has a double seat to insure a tight seal when fully opened or fully closed. The iron safety cap provided to protect the valve should always be in place when the cylinder is not in use.

Oxygen cylinders are tested to withstand twice the normally required pressure. In addition, a safety device is fitted io the valve of each cylinder to relieve the oxygen in case the pressure in the cylinder should get out of hand. When the temperature reaches 208° to 220° F., and pressure builds up between 2,700 and 3,000 pounds, the safety device breaks and releases the oxygen, thus preventing any dangerous pressures within the cylinder. This safety device consists of a thin, frangible metal alloy disc and a fusible plug arrangement.

ACETYLENE

Acetylene gas, a compound of carbon and hydrogen, is the fuel used in the production of the high temperature (5,700° to 6,300° F.) oxy-acetylene flame. It is a colorless gas with a disagreeable odor derived from the calcium carbide from which it is made. When carbide is dropped into water, bubbles of acetylene gas arise which, if lighted, burn with a very smoky flame.

Mixed with air or oxygen, acetylene forms an extremely explosive mixture. A spark introduced into such a mixture will cause almost instantaneous combustion and a violent explosion. The range of explosive mixtures varies from 3 percent acetylene and 97 percent air to 82 percent acetylene and 18 percent air.

ACETYLENE CYLINDERS

Navy standard acetylene cylinders are constructed of a seamless shell with welded ends and are built to hold 225 cubic feet of gas. They are approximately 12 inches in diameter and 36 inches long. The Navy color marking for acetylene cylinders is yellow.

Acetylene cylinders are quite different in construction from oxygen cylinders. Because free acetylene should not be

compressed above 15 p. s. i., the steel cylinders are packed with a porous material. The fine pores of this material are then filled with acetone, a liquid chemical capable of dissolving or absorbing many times its own volume of acetylene.

The cylinders are provided with a valve having a threaded outlet connection for attaching an acetylene regulator. Safety fuse plugs are attached at each end of the cylinders to meet any fire emergencies (see fig. 7). These safety plugs melt at a temperature of 212° F., allowing the gas to escape from the overheated cylinder.

REGULATORS

Regulators perform the important functions of reducing the pressure and controlling the amount of gas flowing from

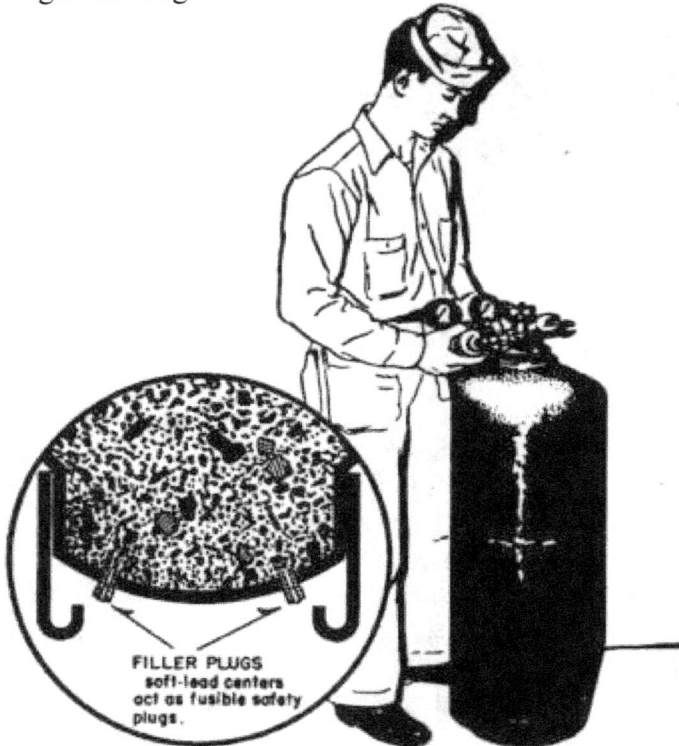

FILLER PLUGS
soft-lead centers
act as fusible safety
plugs.

Figure 7.— Attaching th* regulator unit to an ocotylono cylinder.

the cylinders to the welding torch. They may be adjusted to reduce the cylinder pressure to any desired point and maintain such pressure constant without further attention.

In order to maintain constant torch pressure—without regard to cylinder pressure—regulators must have a sensitive regulating mechanism in addition to a reducing valve.

A regulator has a union nipple for attaching to the cylinder and an outlet connection for the hose leading to the torch. These devices are equipped with two pressure gages-a high-pressure gage showing the pressure of gas in the cylinder, and a low-pressure gage indicating the pressure of gases flowing to the torch. The working pressure is adjusted by means of a handscrew. Changes in pressure are made simply by turning the handle until the desired pressure is registered.

The high-pressure gage on the oxygen regulator is graduated in pounds per square inch from zero to 2,000, and the low-pressure gage is usually graduated in pounds per square inch from zero to 100 or from zero to 150.

Acetylene regulators are not made to withstand the high pressures common to the oxygen regulators, although the two are practically identical in design. On an acetylene regulator, the high-pressure gage is capable of registering pressure up to 50 p. s. i., and the working gage may also register as high as 50 p. s. i.

An oxygen regulator can be either a two-stage or a single-stage type. The two-stage type is preferable when using a portable welding outfit. The single-stage type may be used at individual welding stations where the stationary type of equipment is installed.

The two-stage oxygen regulator has a dual action. When the cylinder valve is opened, the regulator automatically reduces the initial cylinder pressure to about 200 p. s. i. Then the pressure-adjusting screw is turned to the right (clockwise) until the required pressure is shown on the working pressure gage.

A typical two-stage oxygen regulator is shown in figure 8. This type of regulator has two independent diaphragms

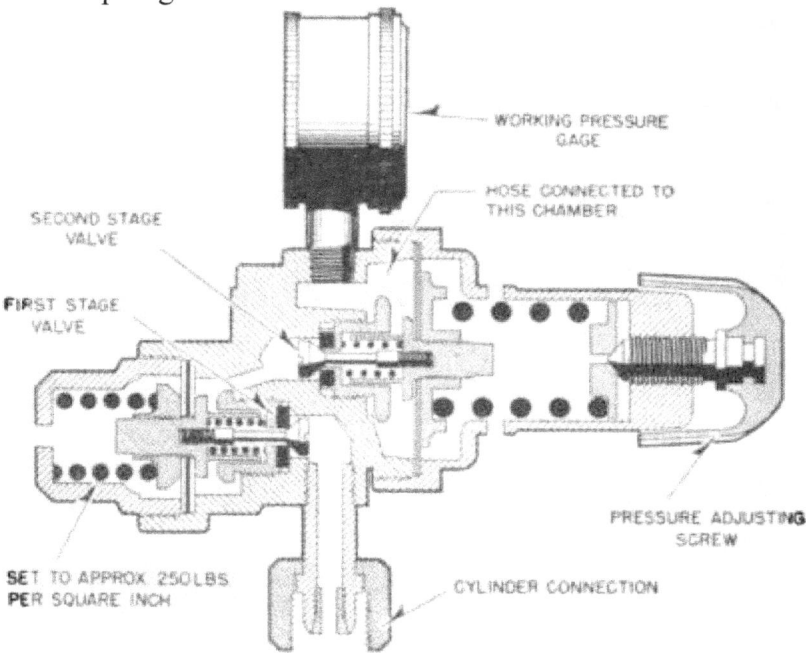

Figure 8.—Sectional view of typical two-stage oxygen regulator.

and stem-type valve assemblies, which render operation extremely efficient. The first stage extra-heavy diaphragm, with its heavy springs, reduces the full cylinder pressure of 2,000 p. s. i. to about 250 pounds. The first stage is wholly automatic and nonadjustable. The second stage differs in that it has a larger diaphragm and lighter springs. A pressure-adjusting screw makes it possible to obtain any desired working pressure.

This second stage is not required to carry the full cylinder pressure load, but functions within a comparatively narrow range. Dual arrangement of valves and diaphragms insures much more constant delivery pressure adjustment than is possible with single-stage regulators.

WELDING TORCHES

Welding torches are devices producing an oxyacetylene flame under conditions of complete control as to flame size,

characteristics, and ease of applications. Welding torches may vary in design, but all types have the same fundamental characteristics.

Welding torches have a handle with two inlet connections for gases at one end, and each is equipped with a valve for control of the volume of gas passing through. Desired proportions of oxygen and acetylene are thus allowed to pass through the torch and thoroughly mix before issuing from the tip or nozzle where the oxyacetylene flame is created by igniting the mixture.

There are two general types of oxyacetylene torches—the equal or balanced-pressure type, and the low-pressure or injector type. The balanced-pressure torch is designed to operate under an acetylene pressure of 1 pound or more. Low-pressure, or injector-type, torches operate with a low acetylene pressure of not more than 1 pound per square inch.

In equal pressure torches, oxygen and acetylene are supplied at independent pressures to a mixing chamber, the construction of which depends upon the particular make of torch. In the injector-type torch, the oxygen at a comparatively high pressure sucks the acetylene into the mixing chamber where the two gases are thoroughly mixed and passed on to the tip.

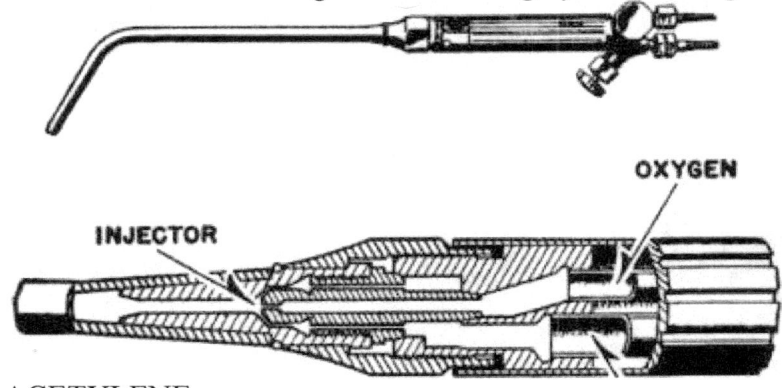

OXYGEN

INJECTOR

ACETYLENE

Figure 9.—Typical welding torch.

238500' -53 3 23

To understand the construction and operation of a welding torch, let us examine the typical type shown in figure 9. At the right of the illustrated unit is the rear bodt which contains the hose connections and the inlet valves for oxygen and acetylene. Attached to the rear body is the handle which is a brass tube with the front body set in the other end to form the terminal for the two smaller tubes inside the handle and to receive the injector located in the rear of the head.

Mixing takes place in the torch head, the rear end of which is shown in cross-section attached to the handle in figure 9. The oxygen passes through the center of the injector nozzle. Surrounding the nozzle are a number of acetylene passageways. As oxygen passes through the small orifice of the injector nozzle, its velocity is increased and a suction is produced that draws acetylene in through the side openings. The passage through which the mixed gases then pass increases in diameter as the stem portion of the head is reached. This increasing diameter permits the gases to expand, insuring thorough mixing. The resultant mixture issues from the torch tip and burns with the proper flame.

Cutting torches are so designed that the flame is fed from several small jets surrounding a central opening in the tip. The central opening—connected by an independent tube and valve to the oxygen supply—provides the jet of oxygen which does the cutting.

TORCH TIPS

Torch tips deliver and control the final flow of gases. They are made of copper because it is an excellent conductor of heat, because it is non-corrosive, and because it has a low coefficient of expansion. The tips, being close to the work and therefore to the heat, must be able to dissipate heat quickly if the welding torch is to function properly.

For some types of torches, heads are available with either a one-piece drawn stem or a two-piece stem, the extra part consisting of a detachable tip. In either case, tips are fastened

to the torch by means of a union nut so that the tip may be adjusted to any convenient angle.

TIP SIZES

There is no standard system for indicating the size of the opening in the torch tip, which determines the amount of heat available for welding. It is important that the correct tip be selected and used with the proper gas pressures if a job is to be satisfactorily welded. If too small a tip is used, the heat provided will not be adequate to secure penetration to the required depth. The use of too large a tip involves the danger of burning or overheating the metal.

Each manufacturer uses a different number system for the various sizes of tips provided. Some comparison of the tips supplied by various torch manufacturers can be had from a study of Table 2 which shows the diameter of the opening, the corresponding drill size, the numbers used by the various companies to designate different tips, and the size of metal for which the tip is adapted. This last item is approximate, as it is based on the tables provided by several companies. Furthermore, the size of tip required is affected not only by the thickness of the metal to be welded, but also by the heat conductivity of the metal.

The best rule to follow is to remember that the smaller numbers designate small gas passages, producing a small flame suitable for welding thin metal, and the higher numbers denote comparatively larger gas passages and a large flame suitable for welding thicker metal.

Use of the correct size tip is not only instrumental in securing a sound weld with good penetration and even overlapping ripples, but also aids materially in eliminating backfires. Experience will enable the welder to automatically select the proper tip for any welding job. The following table is included principally for the information of the newcomer to aircraft welding, as a ready source of reference pending the acquisition of experience.

TobJ. 2.—TORCH TIP SIZES

Tips are provided with protective dust caps which should be left on when the tips are not in use. Remove the dust cap before installing the tip and wipe the threads and seats clean so that no dust or foreign substances remain to prevent the tip from fitting tightly onto the torch head. Tips should be screwed on with the fingers and then tightened finally with the proper size open-end wrench. Pliers should never be

USED TO REMOVE OR TIGHTEN A TCP.

Frequent use of the torch will create a formation of carbon in the passages of the tip. Holding a torch closer than necessary to the molten puddle may permit particles of molten

metal to pop into the tip. Clogged tubes and tips are indicated when greater pressure of the gases is necessary to produce the flame than is normally required for the size of tip being used, causing the flame to be split or distorted.

Tips that have become clogged from carbon deposits or by foreign matter should be properly cleaned out before use. Clean the tip, preferably with the proper size tip drill, or with a soft copper or brass wire. Fine steel wool may be used to remove the oxides from the outside of a tip.

WELDING FLAMES

In welding with the oxyacetylene process, the source of heat is a flame. If the two gases which produce this flame— oxygen and acetylene—are brought together in the proper proportions, the flame conditions may be corrected for any welding operation.

Welding equipment is designed to enable the operator to produce a flame of the size and character best suited for the job at hand. For most metals, the supply of gas to the torch tip is adjusted so that there will be just enough of each gas to produce a balanced mixture. As a result, the flame produced will be neutral. There will be no unconsumed oxygen to oxidize the hot metal of the weld; neither will there be any unburned carbon (from the acetylene) to be absorbed by the hot metal.

Besides the neutral flame, there are two other basic flames that can be obtained from the oxyacetylene torch— carburiz-ing and oxidizing. These flames can be created by changing the relative proportions of acetylene and oxygen. Changing the proportions of the gases directly affects the chemical characteristics of the flame, and in turn determines the action of the molten metal under the flame.

NEUTRAL FLAME

The neutral flame is the one best suited to the welding of most metals since it gives the hottest possible flame with no

NEUTRAL FLAME
BALANCED MIXTURE . Brilliant white cone surrounded by larger
"envelope flame" of pale blue color.

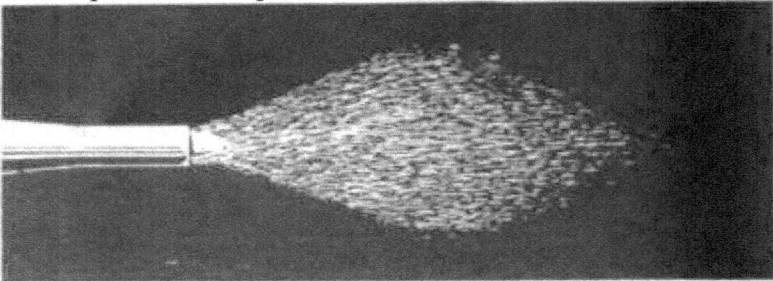

OXIDIZING FLAME EXCESSIVE OXYGEN. Similar to neutral flame; shorter, neck-

in,

and acquires a purplish tinge.

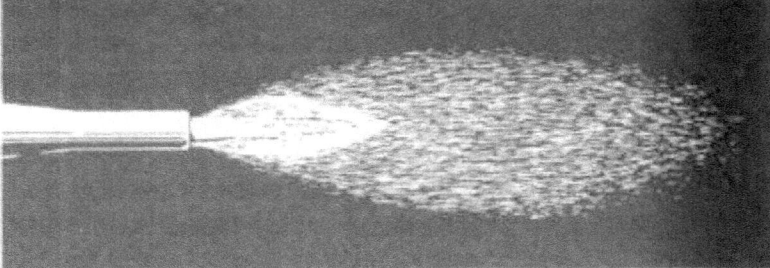

CARBURIZING FLAME
EXCESSIVE ACETYLENE. Three distinct zones. Brilliant white inner
cone, whitish intermediate cone, and bluish outer envelope.
Figure 10.—Types of welding flames.

excess carbon or acetylene present. The neutral flame has a characteristic appearance, featured by two sharply defined zones. The inner portion of the flame consists of a brilliant white cone from i/ 16 -inch to %-inch long. Surrounding this is a larger cone, or "envelope flame," only faintly luminous and of a delicate bluish color.

In the neutral flame, two and one-half volumes of oxygen are required to burn one volume of acetylene. The air surrounding the flame supplies one and one-half volumes of oxygen, the remaining volume being supplied through the torch. Since the proportions of oxygen and acetylene supplied by the torch are equal or balanced, the flame produced by this mixture is called a neutral flame.

The neutral flame is of particular importance to the welder because it is used for such a wide variety of welding and cutting operations and because it serves as a basis of reference in making other flame adjustments. Therefore, one of the first duties of a welder should be to become perfectly familiar with the appearance and characteristics of the neutral oxy-acetylene flame.

CARBURIZING FLAME

A flame with an excess of acetylene is known as a reducing or CARBURiziNG flame because the unburned carbon in the flame is readily absorbed by the molten metal, producing a brittle, carburized surface.

The carburizing flame consists of three easily recognizable zones instead of the two existing in the neutral flame. There is still a sharply defined inner cone and the bluish outer envelope, but between these—surrounding the inner cone— is an intermediate cone of whitish color, as may be seen in figure 10. The length of this intermediate or excess acetylene cone may be taken as a measure of the amount of excess acetylene in the flame.

OXIDIZING FLAME

If there is an excess of oxygen in the flame mixture, the flame is known as an oxidizing flame. It has the general

appearance of the neutral flame, but the inner cone is shorter, slightly pointed, and tinged with a bluish color. It has a tendency to form oxides in the hot weld metal, causing a weak and porous mold.

A slightly oxidizing flame is used in bronze welding and bronze surfacing, while a more strongly oxidizing flame is used in fusion welding brass and bronze. An oxidizing flame can be recognized when welding ferrous metals by the numerous sparks which are thrown off as the metal melts and by the white foam or scum which forms on the surface. An oxidizing flame is harmful to ferrous metals and is carefully avoided on that account.

FLAME ADJUSTMENT

In order to become familiar with the characteristics of the various types of flame and with the adjustments necessary to obtain them, light the welding torch with the acetylene valve wide open and the oxygen valve just slightly open. The acetylene will burn with a smoky yellow flame and will give off quantities of fine black soot.

Now open the torch oxygen valve slowly. The flame will gradually change in color from yellow to blue and will show the characteristics of the excess acetylene flame—that is, there will be three distinct parts to the flame; a brilliant but feathery-edged inner cone surrounded by a secondary cone, and a bluish outer envelope forming a third zone.

With most torches, there will still be a slight excess of acetylene when the oxygen and acetylene valves are wide open and the recommended pressures are being used. Now close the acetylene valve of the torch very slowly. It will be noticed that the secondary cone gets smaller until it finally disappears completely. Just at this point of complete disappearance the neutral flame is formed.

In order to see the effect of an excess of oxygen, close the acetylene valve still further. A change will be noted, although it is by no means as sharply denned as that between the neutral and excess acetylene flames. The entire flame will decrease in size and the inner cone will become much less sharply defined.

ADJUSTING TO A NEUTRAL FLAME

Because of the difficulty of making a distinction between the excess oxygen and neutral flames, an adjustment of the flame to neutral should always be made from the excess acetylene side. Always adjust the flame first so that it shows the secondary cone characteristic of excess acetylene; then, increase the flow of oxygen until this secondary cone just disappears.

During the actual welding operations, where a neutral flame is essential, the flame should occasionally be checked to make certain it is neutral. This is accomplished by momentarily withdrawing the torch from the work and increasing the amount of acetylene until a distinctive feathery edge appears on the inner cone. The amount of acetylene is then slowly decreased until a well-defined cone, characteristic of the netural flame, is formed.

With each size of tip, a neutral, oxidizing, or carburizing flame can be obtained. It is also possible to obtain a hard or soft flame by increasing or decreasing the pressure of both gases.

HOSE

Hose, made especially for the purpose, is used to connect the welding torch to the oxygen and acetylene cylinders. This hose is nonporous, durable, strong, and as lightweight as practicable. The Navy requires welding hose to withstand a hydrostatic pressure test of 250 pounds per square inch.

Red or maroon hose is used for acetylene, and green or black for oxygen. In addition, the name of the gas is usually printed on the hose.

At each end, the hose is equipped with a connection by which it is attached to the regulator outlet and torch inlet.

A connection with a right-hand thread is used on the oxygen hose, while a left-hand nut is used on the hose carrying acetylene. The left-hand acetylene nut is marked with a groove around its center, as illustrated in figure 11, while the right-

OXYGEN ACETYLENE Figure 11.—Hot* connections.

hand nut for oxygen is plain. These connections are, in addition, marked std. oxy, and std. acet.

LIGHTERS

Always use a friction lighter to ignite the torch. Never use matches. When lighting a torch with a match, the hand must be held so close that it may be burned when the gases ignite.

FILLER RODS

Successful welding by the oxyacetylene process depends to a great extent upon the selection of the proper type of welding, or filler rod. This material not only supplies necessary reinforcement to a weld area, but also adds desired properties to the finished weld or bead. By the selection of a suitable rod, either tensile strength or ductility can be secured in a weld, or both can be secured to a reasonably high degree. Similarly, rods can be selected which will help retain the desired amount of corrosion resistance. In some cases, a suitable rod with a lower melting point will eliminate possible cracks from expansion and contraction. In either case, these desired properties can be secured at the same time as a weld free of holes or oxides is developed.

In the early days of welding, it was assumed that ideal results in welding could be obtained by using filler rods or strips practically identical to the base metal. It soon became evident, however, that the heat of welding developed a cast structure in the weld metal, which in some cases was entirely dissimilar to the original metal.

The rods used for welding of steel indicate some of the developments that came about as a result of the scientific study of the reaction of metals to the welding flame. The first step taken to improve welding rods for steel was to use a commercially pure iron, such as Swedish or Norway iron which contains 0.06 percent carbon or less. Without other elements, this rod produced a weld subject to corrosion.

In recent years, manufacturers have developed many types of rods—such as carbon steel rods and alloy steel rods—most of which contain materials that deoxidize the weld, and remove impurities without the use of a flux. Silicon and manganese are commonly used for this purpose. Similar improvements in nonferrous rods have made possible sound welds whose properties can be largely determined in advance. Copper-coated rods have been developed to prevent the corrosion of the rods themselves.

Welding rods may be classified as ferkous or nonferrous* The ferrous rods include carbon and alloy steel rods as well as cast iron rods, while nonferrous rods include brazing and bronze rods, aluminum and aluminum alloy rods, copper rods, and silver rods. Usually, acetylene welding on ordinary steel with steel filler rods does not require the use of a flux. Welding of nonferrous metal with nonferrous rods, or the brazing of iron and steel with bronze rods, does require the use of flux. Stainless steel is also welded with the aid of a flux to prevent oxidation of the hot metal.

FERROUS RODS

While there are a number of rods available for the welding of steel and iron, the average welding shop may employ only

a few since each manufacturer has some all-purpose rods which can be used in a number of applications. If there is any doubt as to the type of rod to use on a given metal, the suggestion of reputable manufacturers should be used and Navy specifications for that metal consulted.

Plain, low-carbon steel rods find the greatest all-round application in welding since they may be used not only for low-carbon steels, but also for welding chrome molybdenum tubing and welding stainless steel to chrome moly steel. Chrome molybdenum steel that is to be heat-treated is welded with a special high-test rod. Low-carbon steel rods generally have less than 0.06 percent carbon, but steel-alloy rods containing less than 0.15 percent carbon are also used for a general-purpose rod.

Another common type of ferrous rod is the high-test rod of higher carbon content for jobs where higher tensile strength is necessary. With this increase in tensile strength must come a decrease in ductility. The amount of manganese in these rods is increased so that the greater amount of carbon is not lost by reaction with iron oxides, and to insure a clean, sound weld. All fittings and joints of carbon steel that must be heat-treated should be welded with high-carbon rods.

Corrosion-resistant steel should be welded with rods similar in alloy to the base metal. If the stainless steel is to be kept at its highest point of corrosion resistance, the welding rod must be of a type that contains small amounts of colom-bium, molybdenum, or titanium. These are used in stainless steel as stabilizing agents—that is, their presence in the steel prevents or restricts the precipitation of carbides at the grain boundaries, which action is the cause of lowered corrosion resistance.

Navy specifications 46R2c and SR43e indicate the process and rod to be used in the welding of 18-8 (18 percent chromium and 8 percent nickel) strainless steel. In an emergency, strips of the base metal will serve as filler rods.

Fluxes are usually employed on stainless steel to prevent oxidation. These fluxes are brushed on in the form of a thin paste, usually to the underside of the seam, although the rod

may be dipped into the paste to help protect the bead. Stirring or puddling with the rod should be avoided.

Cast-iron rod is used for the welding of gray cast-iron maintenance tools and materials. The composition is similar to the base metal, and, in addition, rather large amounts of silicon are included. A typical specification includes 3.00 to 3.50 percent carbon, 3.00 to 3.50 percent silicon, 0.50 to 0.75 percent manganese, and 0.50 to 0.75 percent phosphorus.

NONFERROUS RODS

Aircraft welding of non ferrous materials consists mostly of the welding of aluminum, and more rarely, of brazing or silver soldering. The rods used for these processes are of primary concern to the Aviation Structural Mechanic

For ordinary shop practice, two types of aluminum rods are used—2S and 43S. The 2S rod is of the same composition as 2S sheet or plate—that is, it consists of approximately 99 percent aluminum. The other rod used is 43S, available either as drawn or cast material, the cast rod being used only on cast aluminum. It is often referred to as the five-percent silicon rod, that being the percentage of silicon which it contains. Silicon produces four desirable reactions. It lowers the melting point; it makes the weld less brittle at welding temperatures; it reduces the amount of shrinkage while cooling; and it makes the weld more ductile. 43S rod is used for welding 53S and 51S and aluminum castings. Frequently, strips of base metal are used for welding in order to make certain that the filler is of the same composition as the metal being welded.

Brazing rods are used for joining ferrous metals in those cases where it is desirable to join them without raising the temperature to the mefting point. Brazing rods are composed of copper, tin, and zinc, a rod of typical composition contains 59 percent copper, 40 percent zinc, and 1 percent tin. Newer rods containing silicon produce greater strength and a more sound weld. Brazing is generally limited to the

repair of machines and equipment flsed in aircraft maintenance.

ROD SELECTION

Welding rods are manufactured generally in 36-inch lengths. Steel rods in diameters from % 2 ~ mcn to %-inch, and cast iron rods from % 6 -inch to ^-inch are procured according to Navy specifications 46R4a and 46R4b.

The size (diameter) of welding rod used is governed by the thickness of the metals being joined. If the rod is too small, it will not conduct heat away from the puddle rapidly enough and a burned weld will result. A rod that is too large will chill the puddle. Since there are other factors that affect the desirable size of rod, the choices shown in Table 3 are suggestive only. As in selecting the proper size welding torch tip, experience will enable you to select the proper diameter welding rod.

TdbU 3.—WELDING HOD SIZES FOR SHEET THICKNESSES

It is poor economy to use anything other than high quality rod, as the risk of failure from inferior rods is not warranted by the slight difference in price. A reputable manufacturer's recommendation for a suitable rod may be accepted since the firm has usually made allowances in rod composition to compensate for any probable changes in filler and base metal caused by the welding process.

SAFETY IN WELDING

The importance of observing safety precautions in the welding shop cannot be too highly stressed. Even though manufacturers have provided every possible safeguard in welding equipment, there still exists the possibility of injury when working with such substances as oxygen and acetylene. The characteristics of each one under pressure, and the high temperature produced by their union, make the observance of safety practices imperative.

■

CLOTHING

The welder's clothing should not provide places where flying sparks or pieces of molten metal can lodge to cause serious burns before they can be removed. Guard against open pockets, turned-up cuffs, or ragged holes in the clothing. Low shoes are also dangerous.

In aircraft welding, gloves are not often used because they make the handling of the light torch and rod rather cumbersome. Gloves are desirable for welding heavy metal or using the cutting torch. Gloves should be kept free from oil or grease, as such substances are likely to burst into flame upon contact with oxygen under pressure.

OPENING CYLINDER VALVES

Before connecting a regulator to a new cylinder, the valve should be opened just enough to blow out any dirt in the valve outlet and then closed immediately to prevent excessive escape of gas. This opening to blow out dirt is known as cracking a valve. Stand to one side of the cylinder when you perform this operation.

Neither oxygen nor acetylene should be used directly from the cylinder. The proper regulator should be connected to reduce the pressure to a safe working volume. When the regulator is attached to the oxygen cylinder, the oxygen cylinder valve is opened very slowly at first to prevent

damage to the regulator or gage. Oxygen valves should be opened by hand pressure only—never by a hammer or wrench. The oxygen cylinder should be opened fully after the pressure gage shows full pressure. The cylinder valve should never be opened until the regulator has been checked to make certain that the pressure regulating screw has been released.

When using the acetylene cylinder, the cylinder valve should be opened slowly about one-half turn and never more than one turn. This is done so that the gas can be cut off quickly in an emergency. For the same reason, the wrench is always left on the valve while the cylinder is in use.

FLASHBACK AND BACKFIRE

Flashback and backfire are terms which are interpreted by many as being synonymous. The fact is that they differ greatly in quite a few respects.

A flashback occurs when the flame disappears from the end of the tip and the gases burn within the torch. This necessitates shutting off one or both of the gases to stop combustion within the torch. Flashbacks are often caused by loose connections, overheating of the torch, and improper pressures at the regulators.

A backfire is defined as a momentary return of the gases, not necessitating shutting off the gases as does a flashback. It is indicated in the torch by a snap or pop, the flame immediately recovering itself and burning at the tip in the usual manner. A backfire is rarely dangerous; however, the molten metal may be splattered when the flame pops. Chief causes for a backfire are: touching the tip against the work, overheating the tip, operating the torch at other than recommended gas pressures, a loose tip or head, or dirt in the seat.

HANDLING CYLINDERS

Both oxygen and acetylene cylinders must be handled with great care since they are charged with gases under high

pressure. Neither should have the valves opened, other than for cracking, until the proper regulators have been attached.

When a cylinder must be moved a short distance, it should be rolled on its bottom while the protector cap is used as a pivot by one hand. When moved greater distances, cylinders should be handled with suitable trucks—not cranes or magnets. Empty cylinders should never be used as rollers or supports. Valves should be kept closed. They should be marked empty and returned to the supplier when the gases are exhausted.

USING OXYGEN

The welder should keep in mind the fact that the characteristics that make oxygen useful are the same factors that make it dangerous. The ability to support combustion is useful when properly controlled, but dangerous if leaking oxygen ignites oil or grease on a cylinder, or anywhere on the welding equipment, oil or grease is never applied

TO ANY PART OF THE OXYGEN EQUIPMENT.

Oxygen helps to produce a useful flame for welding only when properly mixed with acetylene in the torch. No opportunity should be provided for it to mix with acetylene anywhere except in the torch. Oxygen cylinders should be stored separately from acetylene cylinders.

Oxygen should always be used for the purpose for which it is intended—never as a substitute for compressed air.

USING ACETYLENE

The proper use of acetylene requires an understanding of its explosive nature. Various mixtures of air and acetylene have been known to explode if ignited.

Acetylene cylinders are used only in an approximately upright position whenever possible to keep acetone from flowing into the lines. Acetylene is never used from a cylinder without first attaching a pressure reducing regulator to the cylinder. Acetylene should never be used from

238500° 53 4

the cylinder faster than one-seventh of its capacity per hour, and free acetylene should not be used at pressures exceeding 15 p. s. i.

GOGGLES

At all times while welding, cutting, or observing welding work, the eyes should be protected by goggles designed for use with the oxyacetylene process. The light from the inner cone of the oxyacetylene flame is itself quite intense, but the hot metal in the section being welded produces a far greater glare. In addition, the eyes are so close to the work that it is advisable to protect them against flying sparks or bits of molten metal that may be splattered about. Goggles also protect the eyes from reflected heat which dries the surface of the eye, causing irritation.

Correct goggle lenses are made of special colored optical glass that minimizes the effect of glare and at the same time permits the operator to see his work clearly. Lenses are available in light, medium, or dark shades.

QUIZ

1. List the equipment necessary for successful oxyacetylene welding.

2. What does oxygen do in the welding operation?

3. In what two ways are oxygen cylinders marked ?

4. What happens when a spark strikes a mixture of acetylene and oxygen?

5. How are acetylene cylinders marked ?

6. What do regulators do?

7. For what is the high pressure gage used ?

8. For what is the low pressure gage used ?

9. What are the maximum cylinder pressures indicated on regulators for acetylene? For oxygen?

10. Does tip size .0210 have a larger opening than tip size .0700?

11. What type of flame would you use in most welding ?

12. Why would you not use a carburizing flame in most welding?

13. Why would you not use an oxidizing flame in most welding ?

14. How can you distinguish between oxygen and acetylene hose?

15. How should the torch be lighted ?

16. What are the differences between flashbacks and backfires?

IT. What precautions must be used in handling cylinders?

18. Should you use grease on welding or oxygen equipment ?

19. Should oxygen be used as a substitute for compressed air ?

20. Before acetylene is used from an acetylene cylinder, what equipment should be attached to the cylinder?

21. When do you use goggles in the welding process ?

CHAPTER 3

FUNDAMENTAL WELDING TECHNIQUES HOLDING THE TORCH

The first technique to master in using the equipment described in chapter 2 is the proper method of holding the welding torch. Two methods are recommended.

When welding light-gage (thin) metal, the hose should be draped over the outside of your wrist, and the torch held as though you were writing with it. For heavier work, a more comfortable grip is that in which the torch is held as you would hold a hammer, with the fingers curled underneath. To prevent fatigue allow the torch to balance in the hand.

WELDING POSITIONS

Whenever possible, it is desirable to make welds with the stock in a flat position. This facilitates puddle control, and

the welder can work longer periods without tiring. These factors are so important that factories often use large jigs, or fixtures, to which the work is clamped, and which can be easily and quickly rotated by hand or by power to any desired angle.

Aircraft repair, however, involves the welding of so many different types of structures that it will quite often be necessary to do the welding in overhead, vertical, or horizontal, positions, as well as in the normal flat position. Regardless of the position in which it is made, the weld must meet the same standards as to bead uniformity, reinforcement, and penetration.

Where the material is accessible from above, welds are made in the flat position, with the rod deposited from above.

Figure 12.—Four basic welding petitions. 44

A horizontal weld is one in which the line of weld is horizontal but the surface of the work is vertical. In a vertical weld, both the line of weld and the surface of the work are vertical. An overhead weld is one in which the filler metal is deposited from the underside of the joint and the face of the joint is approximately horizontal.

CONTROL OF PUDDLE

Welding in positions other than the flat is complicated chiefly by the fact that the molten puddle, affected by the pull of gravity, tends to seek a lower level. Fortunately, there are a number of forces tending to counteract the pull of gravity. These forces, operating in varying degrees in different positions, are: (1) cohesion of the puddle, (2) support provided by the weld metal (which has cooled) and by the base metal, (3) pressure of the flame against the end of the puddle, and (4) manipulation of the welding rod in the molten metaL Cohesion of the metal is directly affected by the heat of the welding flame. Too great an amount of heat makes the puddle more fluid, and therefore more likely to run.

HORIZONTAL WELDING

Practice in welding plate in different positions will develop the skill necessary to make welds in tubing since the welding of a single joint in tubing will often involve welding in several positions.

Welding in a horizontal position involves the carrying of the bead from right to left (for the right-handed welder). In making a practice weld on plate, the pieces should be tacked while in the flat position, then either clamped in a jig or tacked to scrap stock so that the surfaces to be welded are held vertical. Adding the rod to the top of the puddle will help prevent the molten puddle from sagging to the lower edge of the bead. If the puddle is to have the greatest possible degree of cohesion, the puddle should not be allowed to become too hot.

VERTICAL WELDING

Welding a vertical seam (which may be from 45 to 90 degrees to the horizontal) is not a great deal more difficult than horizontal welding. Vertical seams are ordinarily welded from the bottom up. Cohesion of the metal is not as effective a factor in combatting the pull of gravity in a vertical weld as in an overhead weld.

In a vertical weld, the pressure exerted by the torch flame must be relied upon, to a great extent, to support the puddle. It is highly important to keep the puddle from becoming hot enough to run. It may be necessary to remove the flame for an instant from the puddle to prevent overheating.

Vertical welds on plate are usually begun at the bottom, the puddle being carried up with a forehand motion. The tip should be inclined from 45 to 60 degrees, the exact angle depending upon the desired balance between correct penetration and control of the puddle. The rod is added from the top and in front of the flame.

OVERHEAD WELDING

In welding overhead, the puddle can be kept from sagging if it is not permitted to form in large drops. The rod is used to control the molten puddle. By constant movement of the rod, the molten metal is placed near the rear of the puddle. Less heat is required in an overhead weld because the heat naturally tends to rise.

TORCH MOTIONS

Two common methods of welding with the oxyacetylene torch are known as forehand and backhand.

The forehand method, used in welding most of the lighter gages of sheetmetal, is that in which the torch is pointed in the direction of travel, away from the completed bead, with the rod added in front of the flame. In this method of welding, the bead is carried from right to left for the right-handed welder.

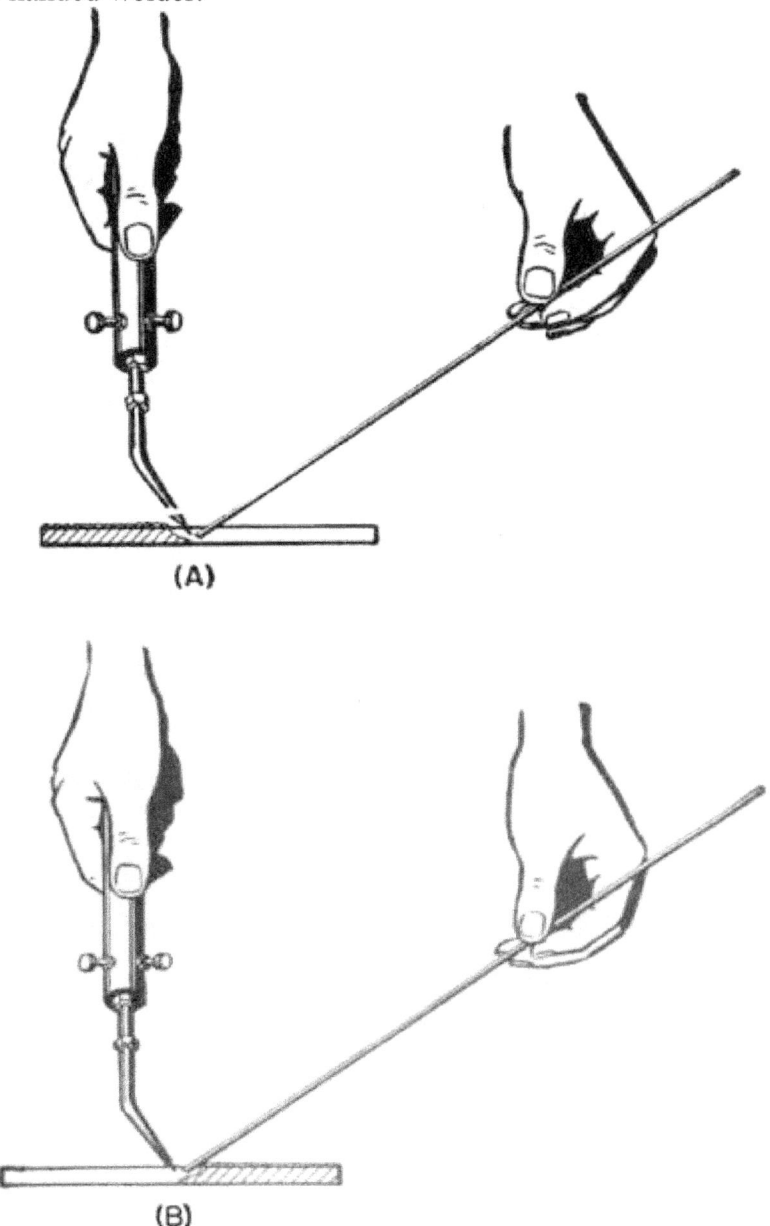

Figure 13.—(A) Forehand welding and (B) backhand welding.

47

The technique of backhand welding is the reverse of forehand welding in most respects. As figure 13 indicates, the torch travel is from left to right (for the right-hand welder), with the flame being directed onto the completed weld. The rod follows the torch and is added between the flame and the finished weld.

An aircraft welder should be equally familiar with both welding methods in order to be able to perform the varied repairs required in aircraft work. Backhand welding is rarely used for sheetmetal welding because the increased amount of heat obtained by this method is likely to cause serious overheating in the lighter gages. However, in the quarter welding of tubing, the backhand method may be of value because of the superior control of the puddle which is possible with this method. This same improved control of the puddle makes it desirable to use backhand welding for heavy stock. Better fusion can be obtained in this heavier metal because it is possible to examine the progress of the weld, and to see that penetration is complete. For these heavier metals, backhand welding is also more economical since less heat is required. Added economy is due to the fact that the more rapid melting of the base metal makes greater welding speeds possible.

BACKHAND TECHNIQUE

The torch is held so that the flame points back toward the completed weld, care being taken to break down the edges and side walls of the base metal, and to fuse them to the required depth. The end of the filler rod is held in the molten puddle between the torch flame and the finished weld.

The rod is given a slight alternating movement toward and away from the flame when adding metal to the weld. This rod movement must be controlled so that the melting metal from the pool will not be pushed over on base metal which is not in a proper state of fusion to receive it.

When performing a weld, bring the torch down until the flame inner cone is about %-inch away from the surface of

the base metal. Hold it there until the flame melts a small puddle of metal, then insert the tip of the rod in this puddle. As the rod melts, gradually work the molten pool forward.

Do not move the torch ahead of the puddle, but work along the edges of the seam slowly enough to give the heat a chance to break down the edges. If the flame is moved ahead too rapidly, the heat will not penetrate deeply enough and the metal will not melt properly. On the other hand, don't be too slow about moving the flame along. If the flame is held in one place too long, the puddle will become too large and it may burn a hole through the metal.

Continue dipping the filler rod in the pool as the torch is advanced. Do not hold the rod so high that the molten metal from the end of the rod will fall drop by drop into the pool. This "raindrop" technique results in a finished weld full of pinholes.

Usually, it is not necessary to put the rod directly under the flame—the heat of the molten puddle will melt it. Keep the flame concentrated on the base metal.

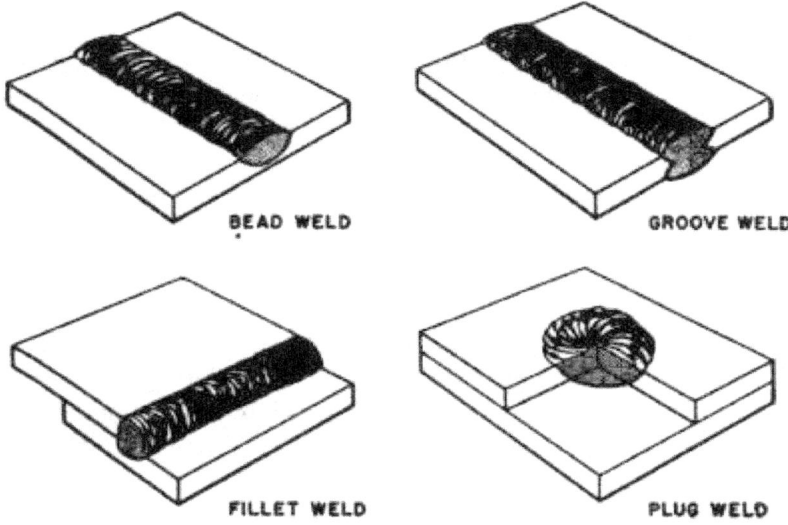

Figure 14.—Four fundamental types of welds.

WELDS AND JOINTS

There are four fundamental types of non-pressure welds—

BEAD WELDS, FILLET WELDS, GROOVE WELDS, and PLUG WELDS.

These types of welds are not necessarily intended to be exactly as shown in figure 14; they are merely representative of good practice.

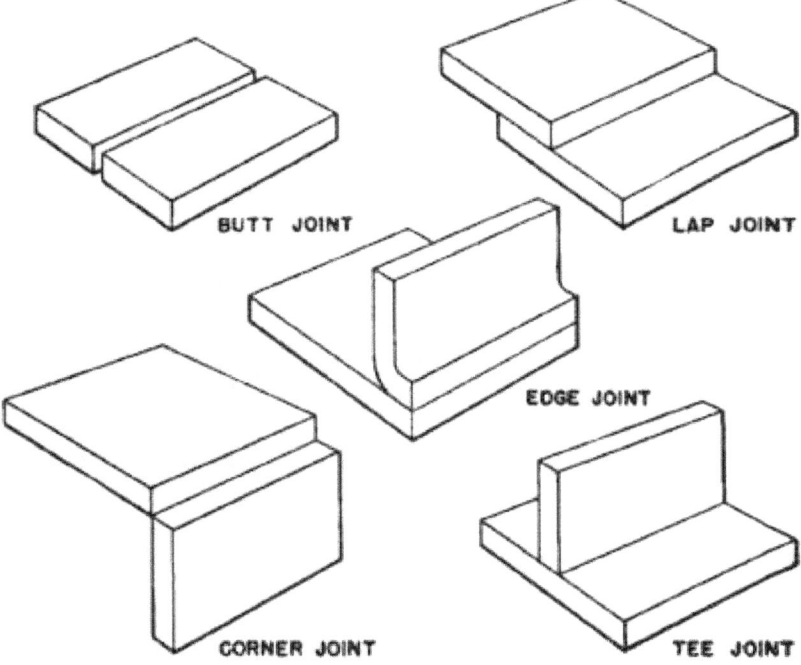

Figure 15.—Five fundamental types of joints.

Five fundamental types of welding joints are butt joint, T-joint, lap joint, corner joint, and edge joint. These are illustrated in figure 15. Most of the welds can be used in various combinations to weld most of the joints, and it is readily seen that many different types of welded joints can result from the combinations of these welds and joints.

LAP JOINT

One of the basic types of welds is the lap joint on sheets, plates, or tubes. The lap weld in sheet or plate is commonly used in spot welding of stainless steel or aluminum as a means of aircraft construction or repair. It is rarely used in oxyacetylene welding of flat stock because of its limitations of strength.

The single lap joint, shown in figure 16, has very little resistance to bending, and will not withstand the shearing stress to which the weld is subjected under tension or compression loads. It is therefore used for sheet, plate, and structural shapes where the load is not severe.

The double lap joint, illustrated in figure 16, has better strength characteristics than the single lap, but it requires nearly twice as much welding as the simpler and more satisfactory butt weld. It is used for sheet and plate parts where the welded sections must have greater strength.

An additional drawback to the use of a lap weld in aircraft ■welding is the added weight. A single lap weld is never used in welding of aluminum because of the difficulty of removing the flux from the joint.

Practice in welding lap joints in sheet or plate is desirable, because it helps you to develop the correct technique for fusing the edge of one part to the surface of another part without building the weld too high or burning away the top edge. This skill will be of value in the welding of structural tubing, or of collector rings and manifolds. The lap joint in tubing is superior to the butt weld, and is used for scarf and fish-mouth splices, reduction splices, finger reinforcements or wrappers, and for silver soldering of fuel lines.

SINGLE LAP

DOUBLE LAP

Figure 16.—Types of lap joints.

Lop Weld Practice Procedure

Overlap the sheets far enough to prevent the annealed zones outside the weld from overlapping. Do not have the metal directly on fire bricks as they will absorb some of the heat from the welding flame.

Tack the plates at V£- to V£-inch intervals (for steel one-sixteenth to one-eighth inch in thickness). After the tacks are made, the pieces should be hammered close together. If contact between the two plates is not maintained, there will be a tendency to burn away the exposed edge.

Weld from right to left if you are a right-handed welder. Direct the cone of the flame slightly more to the bottom plate, assuming the weld is made in a flat position. Add the rod at the top so as to keep the top edge from being cut back and to permit the heat to penetrate to the bottom of the weld. The formation of the correctly shaped bead is an indication that the heat is being properly directed. See figure 17.

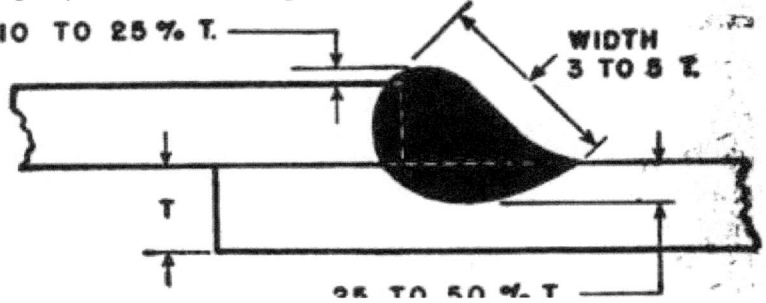

25 TO 50 f* T. Figure 17.—Lap wdd.

A good lap weld has a fusion zone which includes 100 percent of the edge of the upper plate (in the flat position), and penetrates to a depth of 25 to 50 percent of the lower plate.

T-JOINTS

Where the end or edge of one piece is to be welded to the surface of another, the joint is called a T-joint. Such

joints are quite common in aircraft construction, especially in tubular structures. Tee joints may also be formed by welding one plate at approximately a right angle to another plate, or by welding tube to plate. These T-joints are illustrated in figure 18.

T-joints in tubing involve special technique in the actual welding, and will be considered in our discussion of aircraft tubing. The welding of a T-joint in plate is a rather simple process if the correct procedure is followed and the necessary precautions observed.

The plain T-joint in plate, which is welded from one side, requires no preparation other than cleaning the end of the vertical member and the surface of the horizontal member. This joint is suitable for most metal thicknesses used in aircraft work. Heavier thicknesses of stock will have the vertical member single or double beveled like the butt joint in order to permit thorough penetration.

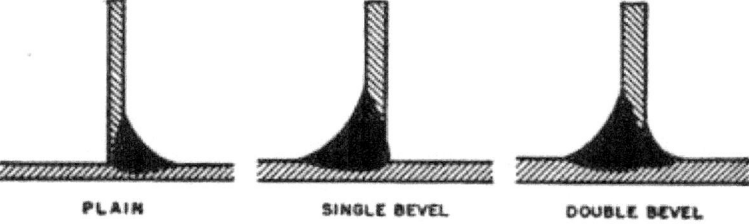

PLAIN   SINGLE BEVEL   DOUBLE BEVEL

Figure 18.—Types of T-joints.

T-Joint Practice Procedure

Put the two pieces of plate in position and tack them near the ends and at other points as required by the specifications for the metal (usually iy 2 to 2 inches). The space between the horizontal and vertical members should not be greater than one thirty-second inch.

Begin welding at one end and proceed toward the opposite end slowly enough to get both pieces in a molten state at the same time. The tip should be directed toward the

root of the weld at approximately a 45° angle, and tilted slightly forward. The rod should be added nearer the vertical member—that is, the rod should be added near the top of the fillet to prevent undercutting. Adding the rod here also permits the full flow of heat into the joint. The inner cone of the flame should be about one-eighth inch away from the surface.

Finish the weld, being careful not to apply too much heat to the vertical plate. If a tendency to undercut the vertical plate persists, play the flame more directly on the piece being undercut.

Precautions

Some results of faulty technique in welding a T-joint are

UNDERCUTTING, POOR PENETRATION, INSUFFICIENT REINFORCEMENT, or too much reinforcement. The undercutting can be avoided by applying the heat equally to both members, and by changing the torch angle as suggested in the preceding paragraph. Sufficient weld metal should be added to obtain a reinforcement in the throat of the fillet weld of 1*4 to V/ 2 times the thickness of the plate. Twenty-five percent of the vertical member at the throat of the weld should be a part of the fused zone; the fused zone should penetrate to 50 percent of the thickness of the horizontal plate. Extreme care must be observed at the start and finish of the weld to avoid overheating of the metal.

CORNER JOINTS

Figure 19 illustrates the different types of corner joints commonly used in making tanks, boxes, and other articles from sheet and plate metals.

The closed type corner joint, shown in (A) of figure 19, is used on thin sheet metals where load stress is secondary. It is made by lapping one piece of stock over the other to fashion a corner. Because the edge of the overlapping sheet is

54

7j

(A)

CLOSED TYPE

1

(B) | OPEN TYPE |

(C)

1

Figure 19.—Type* of corner joint*.

melted down and fused to make the weld, little or no filler rod is added to the closed corner joint.

The open joint, as seen in (B) of figure 19, is used on heavier sheetmetal. The two edges are fused together and sufficient filler rod is added to form a well-rounded bead of weld metal on the outside. If such an open joint must bear a fairly heavy load, a weld is also made on the inside corner to give it greater strength, as demonstrated in (C).

BUTT JOINTS

Figure 20 shows four methods of preparing a butt joint— that is, a joint in which the parts are joined end to end without overlapping. The flanged butt joint, in figure 20 is

y///////////,.

W///////////,,.

FLANGED

PLAIN

SINGLE BEVEL DOUBLE BEVEL

Figure 20.—Types of butt joint*.

238500° S3 5

55

used for very thin sheetmetal up to y 16 -inch in thickness. The edges are prepared for welding by simply turning up a flange. The upstanding portion of the flange should extend above the upper surface of the sheet a distance equal to the thickness of the sheet. Flange welds are usually made without adding welding rod.

Hie plain butt joint can be used for metals from y 16 - to %-inch in thickness. A filler rod is used in making this joint in order to obtain a strong weld.

For metals thicker than %-inch, it is necessary to bevel the edges so that the heat from the torch can penetrate completely through the base metal A filler rod is also used for these welds.

Edge welds, as presented in figure 21, are excellent for fittings made up of two or more pieces of sheetmetal where the edges must be fastened together, and where load stresses are not important

If the sheetmetal is thin, use the plain type in figure 21. That shown in (B) is better for heavier sections. Type" (A) requires no filler rod, since on the thin material sufficient metal can be melted down to fill the seam while furnishing adequate reinforcement

EDGE JOINTS

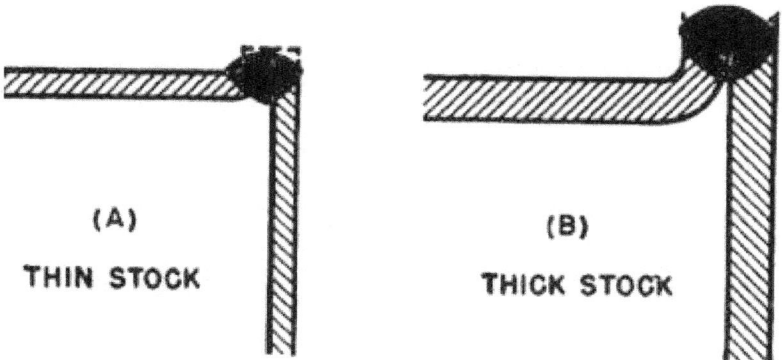

(A)

THIN STOCK

(B)

THICK STOCK

Figure 21.—Type* of edge feints. 56

However, (B) of figure 21 is another story. Here the parts are thicker, and the edges must be beveled or grooved, and filler metal from a welding rod added in order to obtain a strong joint.

Any weld which joins two parts that are at right angles to each other is known as a fillet weld —thus edge joints, lap joints, and corner joints all require fillet welds.

PARTS OF A WELD

Having learned the types of joints used in welding, our next step is to put this information to use. Let us first consider the parts of a weld, each of which has a name, as may be seen in figure 22.

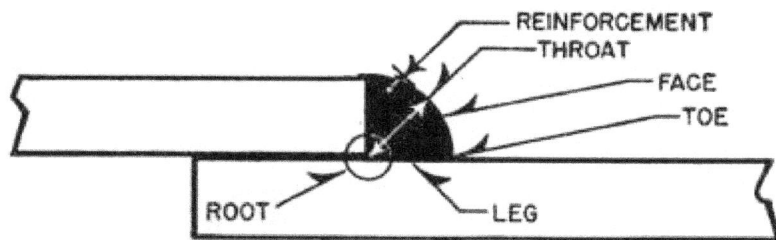

Figure 22.—Parti of a weld.

Parts of a weld are the different areas of the weld metal and the base metal as related to each other in a welded joint. The term "weld metal" refers to the metal in a solidified weld. Base metal denotes the "parent" metal, or the metal in the parts joined as opposed to the metal added to secure the joint in welding. The terms for these areas may be classified as follows:

The face is the top or outer surface of a weld.

The root of a weld is that part at the base of the joint

The reinforcement is the amount of weld metal added above the surface of the base metal.

The throat of a weld is the distance from the root to the face, including the reinforcement.

The toe or a weld is the edge or junction where the face of the weld and the base metal meet.

The leg of a weld (fillet weld only) is one of the fusion surfaces, being the distance from the point where the original surfaces of the parts being welded intersect, to the toe of the weld. A fillet weld has two legs, as the lower drawing in figure 22 illustrates.

APPEARANCE OF WELD

One of the best standards by which to judge the quality of a weld, and the one most commonly used in aircraft work, is the appearance of the weld. The smoothness and regularity of the bead, the presence of fused-in edges along the seam, the amount of reinforcement, and the appearance of the weld opposite the bead are fairly reliable indications of the soundness of a weld.

A properly designed welded joint is stronger than the base metal which it joins. The characteristics of a welded joint that has been properly made are discussed in the following paragraphs.

REINFORCEMENT AND BEAD

Reinforcement of a weld, as we have seen, is the amount of filler material that is built up above the surface of the base metal. The metal that is deposited during the welding process is referred to as the bead.

It is obvious that reinforcement and bead are so closely related that specifications for one directly affect the other. The reinforcement added varies according to the type of weld—butt, lap. or fillet—and according to the thickness of the metal. Sufficient rod must be added to a butt joint to build up the reinforcement 25 to 50 percent of the thickness of the base metal, or the tensile strength of the joint will be inadequate for the loads applied. The bead should be wide enough to tie in or fuse the weld metal with the base metal on each side of the joint. The bead should merge smoothly with

the surface, as shown in figure 23, without undercutting or excessive build-up. .

Adding too much rod to the puddle is likely to cause the filler metal to roll over onto the solid metal past the fusion line of the puddle. The recommended specifications for butt, lap, and fillet welds are shown in connection with the requirements for penetration in these welds.

PENETRATION

Penetration is the depth of fusion in a weld. Thorough fusion is the most important characteristic which contributes

CORRECT PENETRATION COLD EDGES

UNDERCUT POOR PENETRATION

Figure 23.—Penetration of weld*. 59

to a sound weld. In the lighter gages of metal, the normal depth of puddle will insure proper penetration. A normal depth of puddle will be secured if the following precautions are observed: (1) leave a small space (one-half the thickness of the metal) between the edges to be joined, (2) use correct heat, and (3) manipulate the torch and rod properly.

Correct torch manipulation involves holding the torch at the angle best suited to the type of joint and to the position in which the job is being welded. It requires moving the torch along the seam rapidly enough to prevent burning holes in the metal, and yet slowly enough to permit fusion to the bottom of the joint.

Penetration is also affected by the size of the rod, and where and how it is added. Too small a rod will not conduct the the heat away from the puddle, and a burned weld will result. Too large a rod chills the puddle before complete fusion has been secured. Some jobs require manipulation of the rod to keep the consistency of the puddle constant, and to keep it from falling.

SPACING AND PENETRATION

Heavier gages of metal—over one-eighth inch—require preparation before actual welding begins to secure penetration of the weld metal to the desired depth. As previously stated, a space between plates equal to one-half the thickness of the base metal is sufficient for sheet metal. Metals more than one-eighth inch and less than one-half inch in thickness must be single beveled to an included angle of 90°. When the metal is more than one-half inch in thickness, it should be double beveled.

To maintain the spacing decided upon, it is necessary to tack weld the plates. The number of tacks to be used will be determined by the thickness and composition of the metal.

SPECIFICATIONS FOR BEAD AND PENETRATION

In a butt weld, the penetration should be 100 percent of the thickness of the base metal. When penetration is complete in

REINFORCEMENT

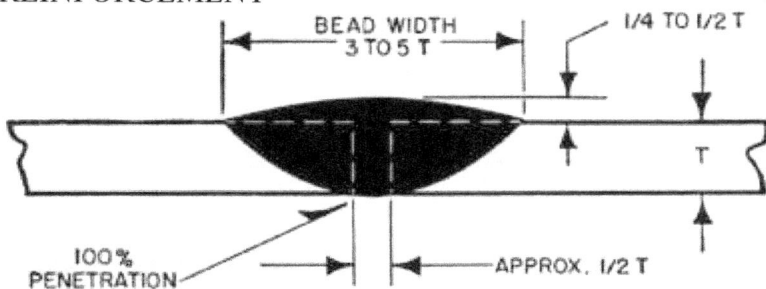

Figure 24.—Butt weld, showing width and depth of bead.

this type of weld, the edges opposite the bead have completely disappeared by fusion, yet no drops of fused metal are evident. The width and depth of bead are shown in figure 24.

On a fillet weld, the penetration requirements are 25 to 50 percent of the thickness of the base metal, as may be seen in figure 25. To determine whether this requirement has been met, examine a cross-section of the weld that has been filed and polished. In this weld, the presence of oxide scale on the side opposite the root of the weld is an indication of the depth of fusion.

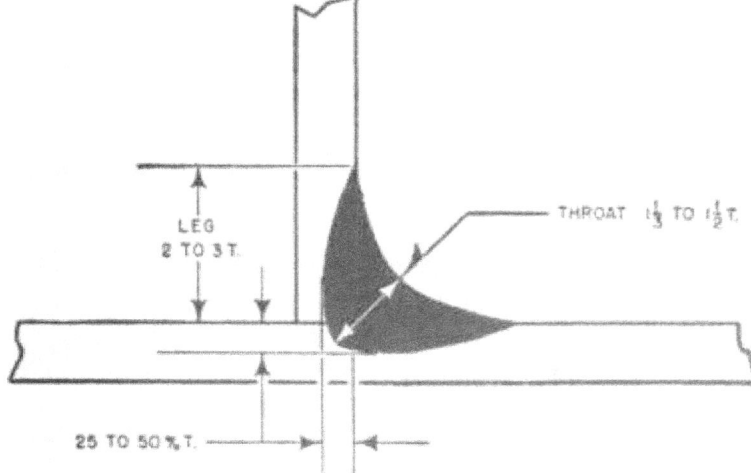

Figure 25.—Fillet weld specifications for bead and penetration.

CHECKING EXPANSION AND CONTRACTION

There are a number of useful methods used in keeping expansion and contraction of metal under control. These methods include tacking the material; employing clamps, jigs, or chill bars; rapid welding; backstep welding; stagger (skip) welding; and, in some cases, preheating the entire piece.

Preheating the entire section of the metal before beginning the welding operation aids in bringing about a uniform contraction when the weld is finished.

Either stagger or backstep welding may be used in rapid welding. Either of these techniques distributes the heat evenly over the length of the seam, resulting in uniform expansion and contraction.

QUIZ

1. What complicates welding in other than flat positions!
2. Where should you start a vertical weld?
3. With what do you control the puddle in overhead welding?
4. What is the forehand method of welding?
5. What is the backhand method of welding ?
6. Make a list of "do not's" for rod and torch techniques.
7. What are the four types of nonpressure welds?
8. What is the objection to a lap weld ?
9. Describe the T-joint.
10. What are the results of faulty welding technique on a T-joint?
11. What is a corner joint?
12. What is a butt joint?
13. When is an edge weld used ?
14. What are the parts of a weld ?
15. How deep should penetration be in a butt weld? In a fillet weld?
16. Name three methods of checking expansion and contraction.

CHAPTER 4

TECHNIQUES FOR FERROUS METALS PRECAUTIONS

The term, ferrous, applies to the group of metals having iron as the principal constituent. Repair by welding of aircraft parts composed of ferrous metals is done only under the following conditions:

1. When the piece to be welded is made of non-heat-treated metal, or, if heat-treated, when facilities are available to reheat-treat.
2. When the metal has not been cold-worked.
3. When the piece has not been brazed or soldered at its joints.

A welder should not repair heat-treated aircraft parts by welding unless facilities are available for reheat-treating. Also, steel parts that depend upon cold-working for their strength are not welded because the heat of the welding flame destroys such strength. Nor should parts made of cold-rolled steel—which include streamline wires, cables, tie rods, and solid drawn wire—be welded.

Steel parts with brazed or soldered joints are never welded by the oxyacetylene method because the brazing or soldering mixture penetrates the hot steel and spoils the weld.

Since heat treatment and cold working are both employed to give metals greater strength, airplane designers make extensive use of heat-treated and cold-worked metal, especially in structural members. Therefore, a welder is correct in deducing from this fact that there are many repair jobs on structural airplane parts that are "out-of-bounds" for his particular skills.

PREPARATION FOR WELDING

Proper preparation for welding is an important factor in every welding operation.

The edges of the parts to be joined must be prepared in accordance with the joint design chosen. The edges must be clean. Arrangements must be made for holding the parts in proper alinement and for preheating, if this is required.

The first step in preparing an aircraft part for welding is to strip it of all dirt, grease or oil, and any protective coating such as cadmium plating, enamel, paint, or varnish. Such coating not only hampers welding, but also mingles with the weld to prevent good fusion.

Cadmium plating can be chemically removed by dipping the edge to be welded in one of the following solutions:

1. A mixture of 73 cubic centimeters of hydrochloric acid, 27 cubic centimeters of water, and 2 grams of antimony trioxide.

2. A mixture of 1 pound of ammonia nitrate and 1 gallon of water.

3. A mixture of 3 gallons of water, 7y 2 gallons of hydrochloric acid, and 1 1 /^ pints of ammonia nitrate.

Enamel, paint, or varnish may be removed by buffing with a wire brush, by the application of emery cloth, by sandblasting, by the use of a paint or varnish solvent, or by treating the piece with a 10 percent caustic soda solution followed with a thorough washing with hot water to remove the solvent and residue.

Sandblasting is the most efficient method of removing rust or scale from steel parts. If dirt, grease, or oil is present, use a one-to-one carbon tetrachloride-naptha solution or a caustic acid solution.

CARBON STEEL

The iron-carbon alloys may be classified as steel if they contain from 0.01 to 1.7 percent carbon, and cast iron if they contain from 1.7 to 4 percent carbon.

Practically all weldable steels in aircraft construction contain less than 0.44 percent carbon. There are a few exceptions, such as the high-carbon fine steel wires which are resistance welded and then braided into cables.

The techniques used in welding low- and medium-carbon steel are more or less basic, although these materials are usually handled in general shop work and not on aircraft. Welding high-carbon steel and alloy steel—for example, stainless steel—requires a similar technique with the addition of certain special precautions.

In general, low- and medium-carbon steels require no preheating. The parts must be thoroughly cleaned, however, and each edge of joints on thick pieces must be beveled down to a 45° angle. A flux is not required. For these steels, a filler rod of low-carbon steel, containing a small percentage of vanadium, is used. The torch flame must be adjusted to neutral, carefully avoiding an oxidizing flame which burns or oxidizes the metal.

GENERAL PROCEDURE

In welding carbon steel, use the forehand method, holding the torch at a 60° angle to the surface of the work. The tip of the inner cone should not quite touch the molten metal.

If the piece of steel is comparatively thick—that is, plate steel rather than sheet steel—use a swinging motion with the torch to make certain that the metal on each side of the groove melts thoroughly. As the torch is swung from side to side of the groove, the edges begin to break down and metal flows in the bottom of the groove.

While this process of breaking down the edges is taking place, the filler rod should be held in the outer cone of the torch flame to heat it. By the time the pool of molten steel has been formed at the bottom of the groove, the filler rod should be almost at the melting point.

As the weld proceeds, filler metal from the rod should be added until the surface of the weld is built up slightly above the edge of the parts. This additional metal provides reinforcement.

The preceding explanation refers to thick pieces of low-or medium-carbon steel. But whether welding thin or thick pieces of steel, it is necessary to critically observe the molten puddle. When it has been built sufficiently high above the surface of the steel, gradually advance the puddle of molten metal along the seam. Be extremely careful, however, not to extend the pool of molten metal until the sides of the groove have been broken down by heat. The reason for this is elementary. Molten filler metal deposited on base metal that is hot but still in a solid state does not fuse. Instead of fusion, the result is adhesion, which is nothing more than soft soldering.

Good fusion is not difficult to obtain, although iron oxide, or scale, in steel melts slightly before the metal does and may be mistaken for molten metal. Remember that a fusion weld is not produced until the metal itself actually melts.

As the end of the seam is approached, raise or tilt the torch flame slightly to chill the molten steel enough to prevent it from flowing over the edge and burning the metal.

The welding principles described above apply equally to high-carbon steel, which owes most of its special physical and mechanical characteristics to its comparatively high-carbon content—at least 0.50 percent. A welding flame with too much oxygen or acetylene will directly affect the characteristics of this steel by changing its carbon content. A carburizing flame (excess acetylene) adds carbon to the weld and makes it hard and brittle. An oxidizing flame (excess oxygen) burns the weld.

High-carbon filler rods aid in maintaining the hardness of steel which is to be heat-treated for increased hardness and strength. Satisfactory results can be obtained with medium-carbon filler rod on thinner sections when considerable intermingling of base metal and filler metal occurs. With the medium-carbon rod, a weld of moderate strength and increased ductility is possible.

STAINLESS STEEL

Steels popularly known as stainless steels are actually corrosion-resisting, although they are not fully resistant to all corrosive agents. Corrosion resistance of stainless steels is determined by the surface condition of the metal, composition of the metal, and temperature and concentration of the corrosive agent.

Corrosion-resistant steel is an alloy steel which includes among its alloying elements chromium or chromium and nickel, which are added to increase corrosion-resisting properties. Slight variations of the chemical composition of these steels produce marked changes in their qualities.

TYPES OF STAINLESS STEEL

Due to their sensitivity to change in chemical composition, a great many types of corrosion resistant steels have been developed, although in general they fall into three classifications or groups:

1. Chromium-nickel steel.
2. Hardenable chromium steel.
3. Nonhardenable chromium steel.

The steels used by the aircraft industry are chiefly of the chromium-nickel group. This group includes those steels containing 17 to 25 percent chromium, 7 to 13 percent nickel, and 0.20 percent or less of carbon. The well-known "18-8" steel (18 percent chromium, 8 percent nickel) is one of this group.

18-8 steel is the corrosion-resistant steel for which the aircraft industry finds the greatest use. It is divided into two groups dependent on their uses in aircraft construction— structural and nonstructural. The structural group has high strength, ease of fabrication, and corrosion resistance, while the nonstructural group is distinguished by its excellent corrosion resistance and heat-resisting qualities such as are required in exhaust collector systems.

The aircraft welder will be concerned only with the nonstructural group of 18-8 steels, for reasons which will be explained in the following paragraphs.

The 18-8 steels in the nonstructural group attain an ultimate tensile strength of 80,000 p. s. i. and an elongation of 40 percent. By comparison, the 18-8 steels in the structural group attain an ultimate tensile strength of 80,000 to 300,000 p. s. i. This great strength is obtained by cold working-such as drawing or rolling—and will be lost if heat is applied. This fact eliminates the possibility of using heat as a means of fabrication and limits its use almost exclusively to parts which can be spot welded.

Stainless steel of the 18-8 nonstructural type may be obtained in sheet form and as welded or seamless tubing.

Welded tubing is less expensive than seamless tubing and is often used in the manufacture of exhaust collectors. Sheet and tubing with a wall thickness of 0.042- to 0.049-inch have proved satisfactory for this purpose.

This material is used primarily for nonstructural parts such as exhaust collectors, stacks, or manifolds. It is practically nonmagnetic in its annealed state and is sometimes used for special purposes, especially in the vicinity of compasses.

All operations necessary for the fabrication of exhaust collectors can be done cold with this steel. The material hardens as it is worked, but it seldom requires annealing before an operation by cold working is completed.

Sheet stock can be bent cold through an angle of 180° without cracking over a radius equal to the thickness of the sheet. Tubing can be bent to an inside radius of two diameters, although a larger radius is preferable.

OXYACETYLENE WELDING OF 18-8 STEEL

The welding equipment used for the welding of ordinary steels can be used for welding nonstructural corrosion-resisting 18-8 steel. The procedure for welding this material is also basically the same as that for ordinary steels. The only difference lies in the fact that certain special precautions must be followed carefully in order to obtain the most desired results.

PRELIMINARIES

Precautions to be taken in welding stainless steel begin with the welding flame. A slightly reducing flame is recommended for use with 18-8 steel. By adjusting the flame so that the feather around the inner cone is about one-sixteenth inch in length, the proper amount of excess acetylene will be obtained.

Protection is thus assured against any variations in gas pressure tending to change a strictly neutral flame into an oxidizing flame. Special care must be exercised to avoid a flame with too much oxygen, as such flame oxidizes the molten metal and renders it porous. Make certain, however, that the flame does not have so much excess acetylene that the stainless steel will be loaded up with carbon and lose its resistance to corrosion.

Use a torch tip one or two sizes smaller than those prescribed for similar gages of plain steel.

The flux should be one especially compounded to dissolve the chromium oxide which forms on the molten stainless steel. Mix the flux with water to form a thin paste. Flux also may be mixed with alcohol or shellac. Apply the flux to the underside of the seam to protect the hot metal from the air and consequent oxidation. Allow the flux to dry for several minutes after it is brushed on so that it becomes fairly solid before the weld is begun.

The filler rod should be of the same composition as the base metal If the base metal is columbium-treated, then a colum-bium-treated stainless steel filler rod should be selected. A filler rod is always required when the pieces to be welded are one-sixteenth inch or thicker.

If the pieces are thin—that is, up to one-sixteenth inch in thickness—the common treatment is to turn up flanges on each edge to a height equal to the metal thickness. The flanges are painted on both top and bottom with flux, then melted down to form a smooth, moderately reinforced weld. In this instance a filler rod is not required, as the flanges furnish sufficient metal to fill the seam.

Tacking is one of the means of lessening warping and distortion by holding the stainless steel in alinement for the welding process. Thin sheets which are to be butt-welded should be tacked at intervals of 1^4 to V/ 2 inches. Tacking may be accomplished by either one of two methods, both of which require the first tack to be placed at the center of the joint. One method requires the placing of the tacks at correct intervals from the central tack toward one end and then tacking from the central tack toward the opposite end. The other method is effected by placing alternate tacks on each

side of the central tack until the ends have been secured. These methods are illustrated in figure 26.

If the parts to be welded are between one-sixteenth and one-eighth inch in thickness, join the edges in a plain butt weld. A backing strip of copper should be placed beneath the seam to prevent the molten metal from flowing out of the weld and

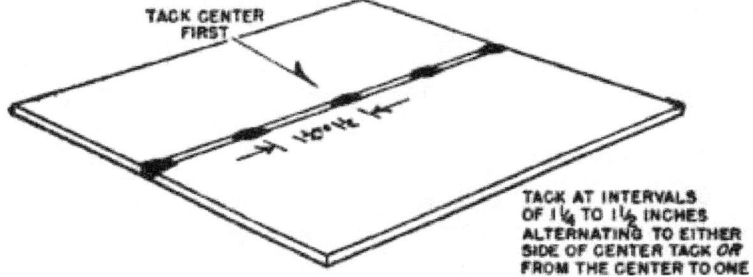

EDGE AND THEN FROM THE CENTER TO THE OTHER EDGE.
Figure 26.—Tacking methods.

to absorb some of the welding heat. If the parts are one-eighth inch or thicker, bevel the edges to provide a V in order to permit heat from the torch flame to penetrate completely through the metal.

CONTROLLING EXPANSION OF 18-8 STEEL

Stainless steel will conduct heat only about 40 percent as rapidly as mild steel, but its coefficient of expansion is about 50 percent greater. These properties cause warping or distortion, especially when the material being welded is thin, unless suitable precautions are taken.

This warping and distortion can be prevented or lessened to a great extent by the use of clamps, copper chill plates, and jigs, or a combination of these appliances to hold the metal in alinement while it is being welded and while the weld is cooling. A precaution to observe when using this method is that the metal must not be clamped too tightly because it tends to buckle and crack if it is not permitted to expand and contract slightly.

238500* 53 —6

Place the parts to be welded so that the line of the weld slants slightly downward in the direction of the welding. This allows the flux, which melts at a lower temperature than the metal, to flow forward and provide protection for the metal as it fuses.

WELDING PROCEDURE

Either the forehand or backhand methods of welding are approved for operations on 18-8 steel, although the former is preferable for thin sheetmetal while the backhand method is recommended for, heavier pieces.

When welding 18-8 steel, puddling of the weld should be prevented or minimized as much as possible because it increases the tendency toward oxidation and separation of valuable constituents.

In the actual welding operation, the flame should be directed forward in order to preheat the metal ahead of the area being welded. The torch is held at an angle of 80° to the surface of the work, with the tip of the flame's inner cone maintained about one-sixteenth inch from the work so that the flame is forced down into the weld. The rod, on the other hand, is held just above the weld and in the flame so that it melts and drops down into the weld as the work progresses.

A relatively small tip is used on the torch to permit slow, careful welding without danger of overheating. Excessive heat on molten columbium-treated 18-8 steel increases the loss of columbium during the welding and should be avoided.

Welds should be completed with thorough penetration in one pass (the layer of weld metal deposited in one trip of the torch and rod down the length of the seam) if at all possible. It is customary and advisable to weld stainless steel entirely from one side. Care should be taken to completely fill the seam so that no point will have to be reworked.

It is important not to stop or to retrace a hot weld. If this is necessary, the weld should cool completely before being reworked. If it should be necessary to go back over a weld,

or if there is need for welding at the back side of the seam in order to support points of severe stress, the entire seam should be preheated before the flame is applied to any local area of the joint. This type of preheating, while necessary, is undesirable because the metal is apt to warp and slow cooling is harmful to stainless steel.

Never start at the edge of a seam and work toward the center. If the joint is in such position that working in from an edge is absolutely necessary, the best policy is to begin the weld at a point an inch or two in from the edge. Weld in from this starting point until that section of the seam is finished, then return to the starting point and weld out toward the edge to complete the weld. This procedure will avoid excessive distortion.

PROPERTIES OF THE WELDED AREA

The welder of stainless steel is concerned not only with the preservation of the strength of the material, but is also responsible for the maintenance of the corrosion resistant quality of the material across the weld zone^ Steels of the 18-8 group are best able to resist corrosion when in their normal austenitic condition. This condition is obtained by heating to approximately 1,900° F., and then quickly cooling to the black (in 3 minutes or less).

In those stainless steels which do not contain columbium or titanium, prolonged exposure to heat in a range from 800° to 1,500° F. will cause the chromium in the steel to combine with the carbon to form carbides, which precipitate along the grain boundaries. These carbides are subject to attack by corrosive agents, such as salt spray or water. Those steels without columbium or titanium are therefore seriously weakened if heated to the range just mentioned and allowed to cool slowly.

BUAER TESTS FOR FERROUS METALS

Before being permitted to weld structural steel parts of naval aircraft, the Bureau of Aeronautics requires a welder
to make the following five types of welded joints on either plain carbon steel or alloy steel.

1. Open single V-butt weld.
2. Open butt weld.
3. Fillet weld.
4. Horizontal fillet weld.
5. Combination sheet and tube cluster fillet weld.

The first joint—the open single V-butt weld, shown in figure 27—tests both the strength of the weld metal and welding technique of the welder. The strength of the weld metal is tested according to standards presented in table 4 which

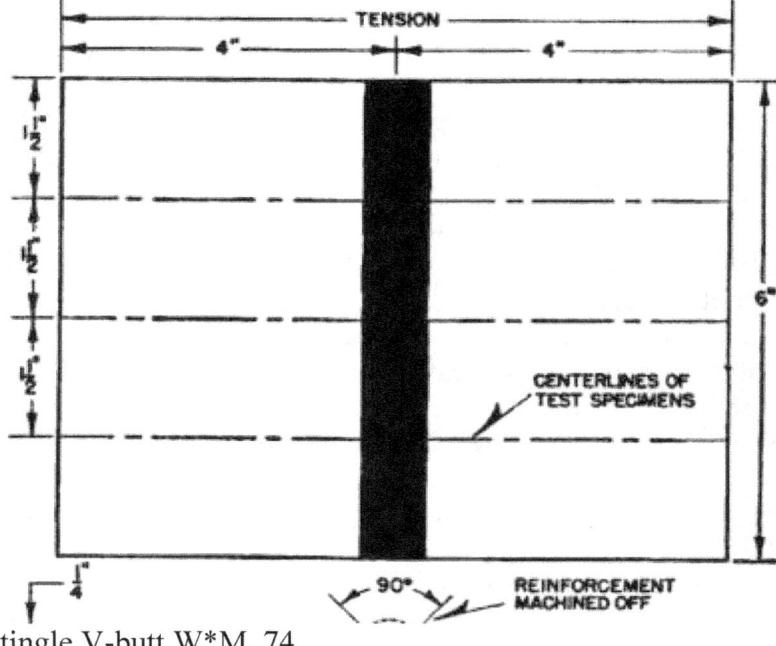

tingle V-butt W*M. 74

shows the tensile strength such weld must have. Welding techniques—whether the metal is warped and to what extent—are based on the appearance of the base metal. -

In making the open single V-butt weld, edges forming the joint are each beveled, usually to 45°, so as to form a 90° V. Weld contour and reinforcement should be as described in figure 27.

The weld meeting the requirements of both tensile strength and welding technique must join two pieces of 14-inch sheet steel spaced one-eighth inch apart. The weld must be built up so that there exists a top reinforcement which can be machined off before the weld is cut up for testing purposes.

Another weld on sheet steel which is part of the test is the fillet weld illustrated in figure 28. This joint is made with the Vi-inch sheets standing on edge during the welding operation. The joint is then tested in tension, and must develop not less than 10,000 pounds per linear inch if the base metal is plain carbon steel, and not less than 15,000 pounds per linear inch if the base metal is alloy steel. Examination of the broken specimen will be made to determine whether or not the weld has fused into the base metal in the toes.

A suitable jig must be improvised for supporting the plates in position while the weld is being made. One such as that illustrated in figure 1 might be used, the only requirements being an ample allowance for clearance between the lower end of the seam and the jig.

Table 4.—TENSILE STRENGTH OF WELD METAL IN POUNDS PER SQUARE INCH 1

« These values are used for alloy-steel filler rod.

Welding operators accustomed to making welds in thin material find the fillet weld the most difficult to manufacture as it requires more skillful preheating of the base metal to obtain fusion in the corners of the welds. It is illustrative of many joints in gear axles and large fittings.

Penetration into the sides of the joint in aircraft welding should be at least one-fourth of the base metal thickness, and also completely through to the root of the weld.

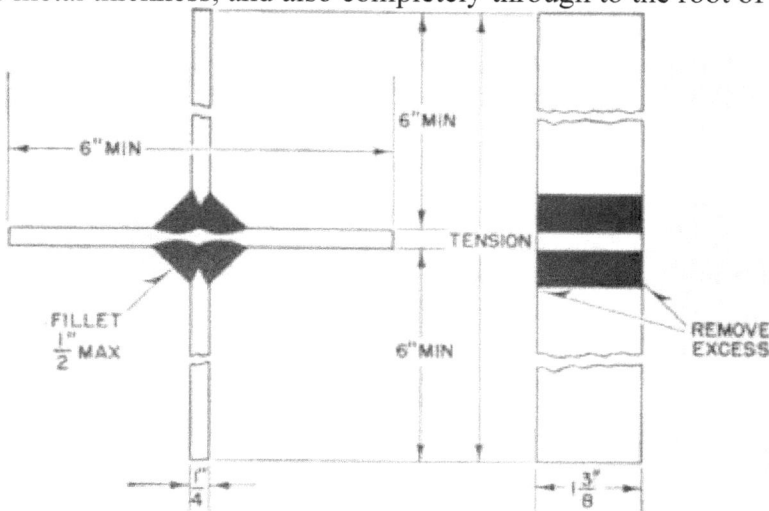

Figure 28.—Fillet weld.

A tubular butt weld, such as presented in figure 29, when performed on tubing of 1-inch diameter and 0.028- to 0.065-inch wall thickness, must have a tensile strength of not less than 50,000 p. s. i. for plain carbon steel, and 80,000 p. s. i. for alloy steel. The reinforcement on the joint may in no case be more than twice the wall thickness of the tubing.

The qualifying welder is required to make the tubular butt weld in three positions—one set, consisting of a horizontal and a vertical specimen, is welded on the bench, and a similar set is welded in an overhead position not lower than the welding operator's eyes.

-TENSION •
BEAD NOT GREATER
5"

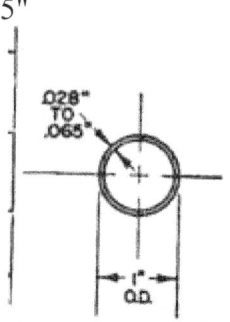

Figure 29.—Tubular butt w«ld.

When welding the tubing in the overhead position, the welder is not permitted to rotate the tubing.

Specifications for a fillet weld made in a horizontal position call for welding plate to tubing. This type of weld, according to the majority of AM's, is the most difficult of the lot to perform. It is not tested for tensile strength, but is rigidly tested for penetration.

Tubing for this test weld must be iy 2 inches in outside diameter, and the plate must be three-sixteenths inch in thickness. Clearance between the plate surface and tubing must be one-sixteenth inch after tacking. The weld must have a % 6 -inch throat and a % 2 -inch leg.

Carefully observe penetration when making this weld, and in addition, have completed and uniform fusion of weld to base metal at all points of contact.

Lack of fusion is not the only failure which will disqualify the welder. The inspector Will be equally impressed by laps, fissures, and gas pockets in the weld.

Another required test weld on steel is a three-tube joint. This joint is a combination sheet and tube fillet weld, and is tested in tension. In this test, three tubes (1 inch in diameter and 0.065 inch in wall thickness) are joined to a piece of sheet metal one-fourth inch in thickness, as shown in figure 30.

The welded seam between the plate and vertical tube must have a minimum tensile strength of 1.500 pounds per linear

inch for plain carbon steel and 2,500 pounds per lineal inch for alloy steel. The welder is also required to obtain full uniform penetration of the weld metal.

After the tension test, the specimen is cut at points designated in figure 30, and then polished and etched to demonstrate the proper and improper methods which govern the penetration of the weld metal into the base metal.

6"MIN

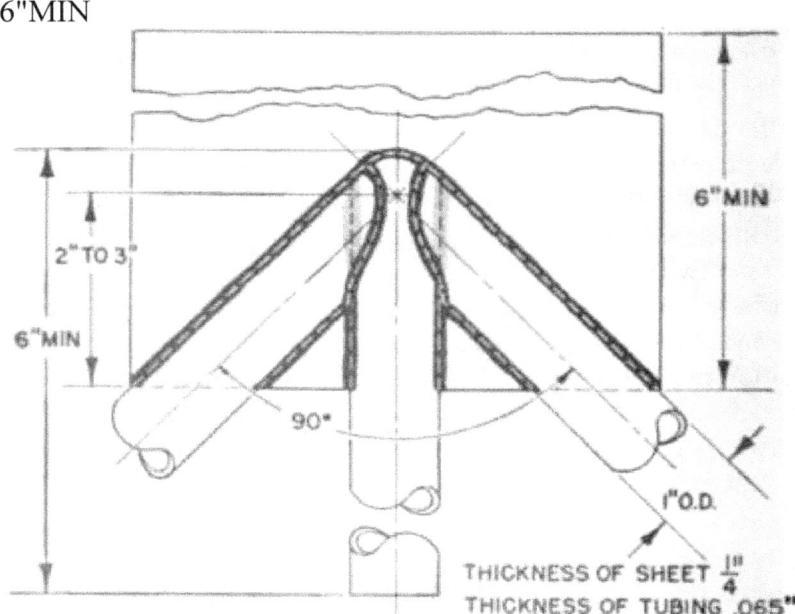

Figure 30.—Three-tube joint.

Because corrosion-resisting and heat-resisting steels require a welding technique slightly different from other types of steel, the Bureau of Aeronautics requires two test welds of the single butt type on both 18-8 stainless steel and nickel-chromium alloy steel.

These welds must join two thin sheets and two heavy sheets in such manner that they will withstand a 180° bend

test. The weld on thin sheets, when bent 180°, must show no evidence of cracking, either in base or weld metal. For the weld on thick sheets, the bend test should produce cracks no longer than one-sixteenth inch in any direction.

The welder also may be tested on welds made on actual aircraft parts composed of 18-8 or of a nickel-chromium steel.

CAST IRON

The welding of cast iron is not a frequent practice in aircraft work, and the inclusion of such technique in this discussion is to familiarize the Aviation Structural Mechanic with the fundamental principles of this process to prepare him for any emergency which might arise and also to enhance his knowledge of all welding techniques.

Gray cast iron—iron containing 2.5 percent carbon—can be either fusion-welded or bronze-welded (brazed). For most cast-iron welding jobs, bronze welding is preferred, although fusion welding is necessary when the part must withstand minimum temperatures of 500° F.

In fusion-welding gray cast iron, four fundamental principles must be borne in mind:

1. Preheat the iron to a dull red before welding.
2. Use the proper flux.
3. Use the correct welding rod.
4. Cool the edges of the weld very slowly.

The edges of a cast-iron piece to be welded must be beveled to form a 90° V. On small parts, this can be done with a hammer and cold chisel, or with a grinding wheel. On large parts, the V is cut with a pneumatic chipping hammer or with the cutting tip of a welding torch.

In beveling ordinary thicknesses of cast iron, extend the V only to within one-eighth inch of the bottom of the break. This blunt bevel makes it easier to control the molten cast iron and lessens the danger of burning a hole through the bottom. Carbon blocks also may be placed under the weld to prevent molten cast iron from running through.

After the joint is beveled, remove all grease, slag, rust, and dirt for an inch back from the edges with an emery wheel, sandblast, wire brush, or cold chisel. Unless the joint is perfectly clean, porous spots and blowholes will exist in the weld.

Preheating the casting insures equalization of the expansion-contraction stresses. Gray cast iron is very brittle and sensitive to changes in temperature. If one part of a casting is heated in welding, while the rest of the piece remains cold, the metal will crack. Preheating the entire piece surmounts this difficulty.

The size and shape of the casting and the location of the break determine whether or not it is possible to confine the preheating operation to the section around the break. Small parts are preheated with the torch flame, while larger pieces should be put in a furnace. In either case, the metal should be heated to a dull red and held at that temperature throughout the welding operation.

The welding rod should be of a special chemical composition which will put certain elements back in the weld metal— particularly silicon—which tend to burn out during the welding.

A flux is also needed to break up the coating of slag which forms on the surface of the molten cast iron. Sprinkle flux on the joint and dip the filler rod into the can of dry flux after the rod has been slightly heated in the torch flame. Flux may thus be introduced gradually into the puddle. A recommended flux for cast iron is one with equal parts of carbonate of soda and bicarbonate of soda.

Use a welding tip one size larger than for steel of the same thickness.

Welding Procedure

When the piece has been cleaned, beveled, and preheated, the next step is to adjust the torch flame to exactly neutral.

Direct the cone of the flame at the bottom of the groove, about one-eighth inch from the surface. When the bottom

so

of the V is thoroughly melted, use a constant, circular motion of the torch to melt the beveled sides of the V until they begin to run downward and combine with the molten puddle at the bottom of the groove. Move the torch constantly from side to side to make certain that both sides, as well as the bottom, are melted. The torch motion in this procedure is much like that used in welding heavy steel plate.

This done, hold the filler rod near the flame to heat it, dip it in the flux, and place it just below the surface of the molten puddle and keep it there, as portrayed in figure 31.

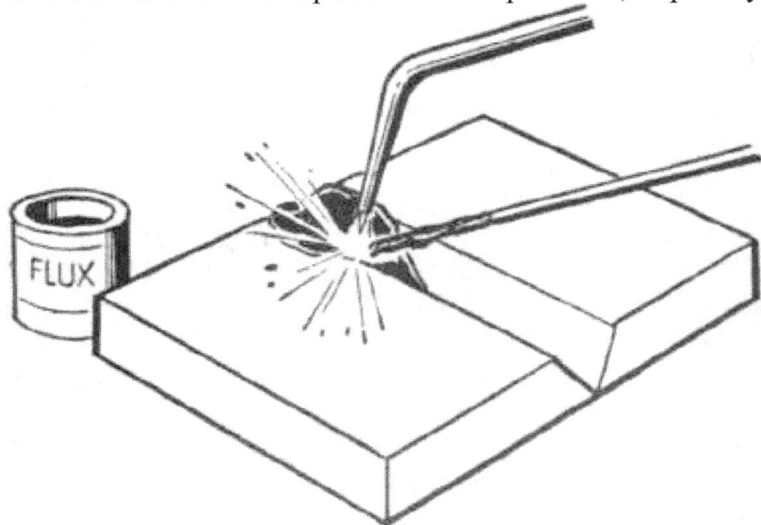

Figure 31.—Cast-iron welding.

Notice that as the filler rod melts into the puddle, the level of the molten metal rises in the groove. Carefully watch this molten puddle, and when it has been built slightly above the top surface of the base metal, move the torch flame forward. Be careful to melt the sides of the V ahead of the advancing puddle so that the molten metal is never forced ahead onto colder metal.

Occasionally stir the molten pool and, with the filler rod, skim off any impurities which may appear on the surface. From time to time add flux to the filler rod.

When gas bubbles or white spots appear in the puddle or at the edges, add flux and play the torch flame around the spot until the impurities float to the top.

Cast-iron welding should be performed as rapidly as possible. When the weld is completed, cover the part with a piece of asbestos paper to enable it to cool very slowly.

QUIZ

1. What are some conditions under which ferrous metals will not be welded?
2. What is the first step in preparing a part for welding I
3. What is 18-8steel?
4. Outline the preliminary procedure for welding 18-8 steel.
5. What is the importance of tacking in welding 18-8 steel?
6. What methods besides tacking should be used to prevent distortion when welding 18-8 steel ?
7. Outline welding procedure on 18-8 steel.

8. What joints are required for Buaer tests?

9. Outline the procedure for welding cast iron.

CHAPTER 5

TECHNIQUES FOR NONFERROUS METALS

AIRCRAFT METALS

In everyday language, nonferrous metal means a metal not made from iron ore. The most important of the non-ferrous metals in aircraft construction are the aluminum alloys. Others are the magnesium alloys, the nickel-based alloys, and inconel and monel, although the latter two are not strictly nonferrous metals.

Aircraft are composed chiefly of these metals because they are strong and light.

Welding of aluminum alloys in modern aircraft repair has been reduced to a very small portion of the total repair required to maintain the plane. Welding, with few exceptions, is limited to the repair of metal fuel and oil tanks, cowling, cast fittings, or cast motor parts in some of the older types of training planes.

ALUMINUM

Commercially pure aluminum is a silver-white, lustrous metal noted for its lightness—it is only about one-third as heavy as steel. It has a high degree of resistance to corrosion and can be readily formed into intricate shapes.

Aluminum is the most widely distributed of all the elements, next to oxygen and silicon. It is present in all common rocks, but is difficult to extract. Aluminum metal is produced by electrolyzing alumina, which is prepared by purifying bauxite—a hydrated oxide of aluminum containing iron oxide and silicon as the principal impurities. Bauxite deposits of satisfactory quality are distributed throughout the world, and alumina of high purity can be obtained cheaply from these deposits.

In order to electrolyze an aluminum compound, it is necessary to dissolve it in a substance which does not react with aluminum and which is more resistant than aluminum oxide to electrolytic decomposition. Cryolite, a double fluoride of sodium and aluminum, is such a substance. Cryolite melts at about 1,830° F. and, when molten, dissolves about 16 percent of its weight of alumina. By electrolyzing a molten bath of cryolite and aluminum oxide with a current of low voltage and high amperage (aluminum oxide is disassociated into aluminum and oxygen) the molten aluminum, being heavier at 1,850° F. than the molten cryolite, settles to the bottom of the bath. Powdered alumina is added from time to time to replace that electrolyzed. The molten metal is removed by tapping and is poured into molds to solidify as pigs. The metal produced has an average purity of 99.5 percent, which is known as commercially pure aluminum.

One virtue not possessed by pure aluminum is that of strength, and aircraft parts must of necessity be strong as well as light. Aluminum manufacturers have met this problem by adding one or more other metals to the pure aluminum to develop alloys which are both light and strong.

These alloys, available in either cast or wrought forms,

can be made still stronger and harder by work-hardening— that is, by rolling, forming, pressing, or otherwise cold-working the metal. Since the hardness depends on the amount of cold work performed, some of the wrought-aluminum alloys are available in several work-hardened tempers—either *4 hard, y 2 hard, % hard, or hard. The y 2 hard temper of aluminum sheet is that most often used for aircraft parts.

To add still greater strength, aluminum alloys are heat-treated, a process of controlled heating and cooling of the metal. With but minor exceptions, aluminum alloys used for structural members in aircraft are heat-treated. The principal aluminum alloy thus used is 24S.

Some of the other heat-treatable alloys used for structural parts of aircraft are A51S, 17S, 53S, 61S, and -25S. These code numbers refer to the kind and amount of alloying metals which are added to the aluminum to give it certain qualities of strength and hardness. The total amount of alloying elements is seldom more than 6 percent in the wrought forms. Cast forms, however, may contain somewhat higher percentages of alloying metals.

Non-heat-treatable alloys used for parts in aircraft which are not subjected to great stress, such as tanks and fuel lines, include 2S, 3D, 4S, and 52S. A cast alloy, 43, is used for many nonstructural fittings.

In general, very little welding of aluminum aircraft parts is done, because the great majority of them are made of heat-treated metal.

An aircraft welder is not permitted to weld structural parts made of heat-treated aluminum alloys. Even if facilities are available for reheat-treating these parts after welding, it is still not possible to increase the strength of the weld enough to stand up to the stresses which these parts must bear. Remember, the use of torch welding in aircraft is limited to places where high unit stresses are not involved and, consequently, where heat-treated metal is not used.

Although the occasions for such practice are infrequent, an Aviation Structural Mechanic must know how to weld aluminum alloys. One common application of torch welding on aircraft aluminum and aluminum alloys is in welding tanks. Torch welding is the simplest method of obtaining gastight or liquidtight seams in such tanks.

GENERAL WELDING PRINCIPLES

In oxyacetylene welding the standard torch, hose, and regulators are suitable for both wrought and cast aluminum workpieces.

As in welding any other metal, the first step is that of cleaning the joint, either mechanically or chemically.

To clean the joint mechanically, use a wire brush, steel wool, or abrasive cloth on the edges to be joined until a dull-white, nonreflecting surface is obtained. Be careful not to scratch the metal beyond the weld area, as such scratches are entry points for cracks and corrosion. If considerable corrosion exists, the joint should be chemically cleaned by dipping it from 10 to 30 seconds in a hot solution of 10 percent caustic soda or 10 percent tri-sodium phosphate, followed by a rinse in a dilute 10 percent nitric acid solution and a final hot-water bath.

After cleaning and otherwise preparing the joint, preheat the piece. A slight preheating of thin sheet is all that is required, and this is accomplished by passing the flame back and forth across the weld area three or four times. Aluminum sheets three-eighths inch or thicker, as well as larger aluminum castings, should be preheated to from 700° to 800° F. When sufficiently preheated, the piece will have lost its metallic ring when struck lightly with another piece of metal.

Preheating avoids heat strains and reduces the amount of welding gas necessary to fuse the seam. Such preheating is imperative because aluminum conducts heat and expands very readily. Castings in particular require careful preheating to prevent serious cracking. SmaH castings may

be preheated with the torch, but larger ones require furnace treatment

Never preheat aluminum alloys to a temperature higher than 800° F., because the heat may melt some of the alloys and result in burned metal.

ROD SELECTION

Choosing the proper filler rod is as important with aluminum as with any other metal. Non-heat-treated 2S and 3S require a 2S filler rod. For 52S, 53S, and 61S, a filler rod containing

95 percent aluminum and 5 percent silicon (43S is the code number) is recommended.

Welding rods are available in % 6 -, % 6 -, and ^4-inch diameters. The usual rule is to match the size rod to the thickness of the base metal.

WELDING TIPS

Since aluminum is an excellent conductor of heat, it is wise to select a welding tip of a size slightly larger than that used for steel of the same thickness to obtain sufficient heat to melt the base metal. Table 5 shows the recommended sizes of tips and amounts of gas pressure for welding aluminum of varying thicknesses.

The welding flame for aluminum work should be adjusted to neutral. Some welders prefer a slight excess of acetylene in the flame, but experience has shown that a neutral flame—one-to-one mixture of oxygen and acetylene—will serve best. The flame should be soft; this requires adjustment of the torch needle valves so that the gas mixture comes into the tip at low speed.

FLUX

Aluminum or aluminum alloys, when exposed to the air for any length of time, will form a thin film of oxide on their surfaces which is troublesome to the welder for two reasons—the oxide has a higher melting point than the metal, and it prevents the free flow of molten metal.

238500° -53 7

TabU 5.—APPROXIMATE SIZE OF TIPS AND RELATIVE OAS PRESSURES USED IN

The most satisfactory method of removing oxides is through the use of a flux, the application of which does several important things. Its chemical action reduces the melting point of the oxide below that of the metal, and it dissolves some of the oxide and floats it to the surface, thus preventing oxide formation.

Flux for aluminum generally comes in powder form, and is usually mixed with water to form a thin paste (two parts of flux to one part water).

Unless the particular job requires a filler rod, paint the flux directly into the joint. If a filler rod is used, coat it with flux. On thick sections of aluminum, both the metal and the filler rod are treated with flux.

JOINT DESIGN

Aluminum may be welded in any of the joint designs; however, butt joint welds on sheet aluminum alloy should be of the flange type, and such joints should be clamped together and tack-welded to hold them in alinement. Tacks are placed at intervals of from 1% to iy 2 inches apart, following the same procedure as for stainless steel. Principles of aluminum butt-joint welding are illustrated in figure 32.

An ordinary butt joint, such as shown in figure 32, may be used when the piece is less than 0.083-inch in thickness. Heavier sheets should be notched. Edges of the joint are

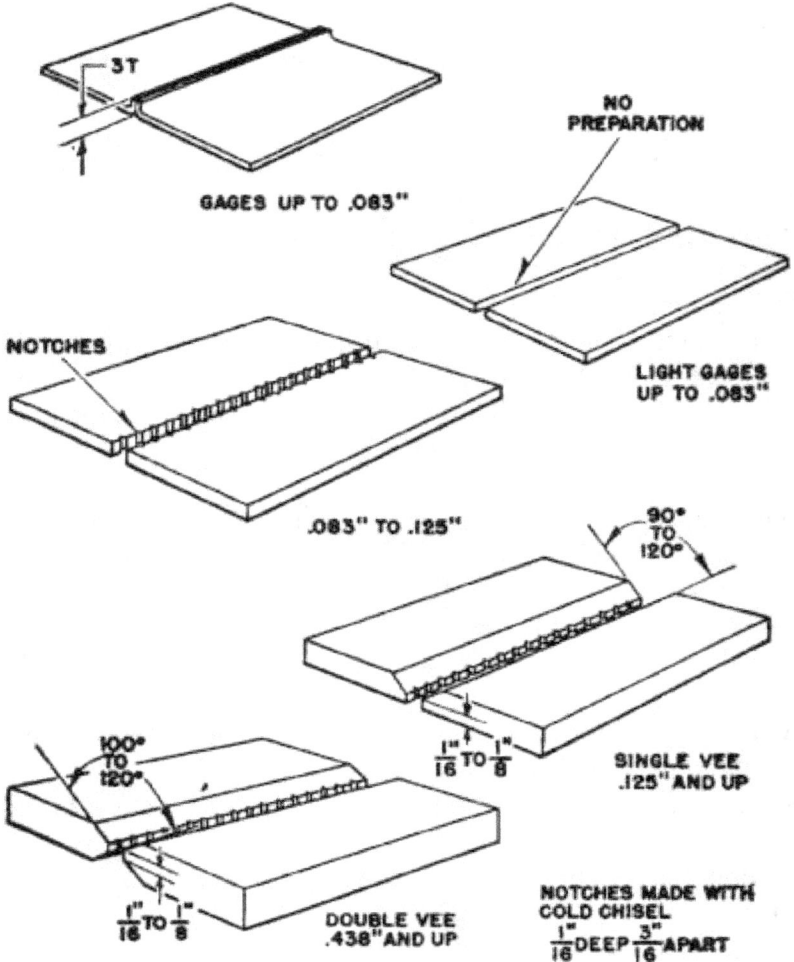

GAGES UP TO .063"

NO PREPARATION

NOTCHES

LIGHT GAGES UP TO .063"

.063" TO .125"

90° TO 120°

$\frac{1"}{16}$ TO $\frac{1"}{8}$

SINGLE VEE .125" AND UP

100° TO 120°

$\frac{1"}{16}$ TO $\frac{1"}{8}$

DOUBLE VEE .438" AND UP

NOTCHES MADE WITH COLD CHISEL
$\frac{1"}{16}$ DEEP $\frac{3"}{16}$ APART

Figure 32.—Butt joints for aluminum welding.

notched with a cold diisel to a depth of about one-sixteenth inch. Spaced three-sixteenth inch apart, these notches act as expansion joints and also aid the flux in thoroughly penetrating into the seam.

WELDING PROCEDURE

The welding of aluminum requires particular attention to a few of the factors involved in the process, the remaining factors being the same as in the welding of other metals. Items requiring this special attention include correct flame adjustment, tacking methods, and torch and rod technique.

Hold the torch at a considerable slant so that it will not blow holes through the metal. An angle of about 45° to the surface of the base metal is just about right for butt joints. In welding a T-joint, the flame is held midway between the two pieces. The inner cone of the flame should be about one-eighth inch from, but never touching, the metal.

Exercise considerable care in evenly heating both edges of the joint so that the heat will be well spread around in the weld area. Do not hold the torch too long in one spot-aluminum conducts heat so well that the entire area around the weld will crumble and fall away.

Another thing to remember is hot short —that is, when aluminum approaches its melting temperature, it loses its power of adhesion and cannot support its own weight. If care is not exercised, the puddle may drop through the metal or cracks may develop from strains set up in the weld area,

If filler rod is used, be careful to see that the filler metal enters the weld only where the base metal has been brought to a molten state.

The problem of deciding exactly when the edges have reached their melting point is complicated, in the case of aluminum welding, by the fact that no visible color change occurs when the melting point is reached. The aluminum may be solid one instant and then, with no change in appearance other than a quick wrinkling and shrinking action, melt and sag.

Aluminum begins to feel soft and elastic just before melting takes place. The instant that this feel indicates approach of the melting point, dip the filler rod into the puddle and permit it to alloy with the metal. A diagram illustrating the dipping motion the filler rod should take as the weld progresses is presented in figure 33.

\S \> \>

(A)

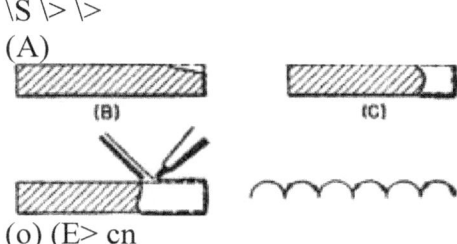

(B)   (C)

(o) (E> cn

Figure 33.—Proper filler rod motion in welding sheet aluminum.

In (A) of figure 33, the metal is being heated; in (Z?), melting of the base metal has begun; (O) shows the filler rod dipped into the molten puddle and allowed to melt; and in the filler rod is lifted while the torch flame is moved forward to continue the melting process. The filler rod dipping procedure is shown being repeated in (E) , and (F) illustrates the continuous dipping motion of the filler rod as the weld progresses along the seam.

In the dipping process described and illustrated above, never lift the rod out of the flame envelope. As the welding process nears the end of the seam, increase the speed of progress to prevent a collapse of the base metal.

In welding sheet or plate, it is an excellent practice to begin near the center of the pieces and weld toward the ends. This procedure better controls expansion and contraction and will prevent, in most instances, the cracking that often occurs when the weld is begun at the very end of a sheet.

Welds on both aluminum sheet and plate should be done with one pass of the welding torch.

Wash all traces of flux from the surfaces of the completed weld. Otherwise, if moisture is present, the elements in the

flux will attack the base metal and cause corrosion. To remove the flux, scrub the piece with a stiff brush and hot water. If the weld is difficult to reach with a brush, submerge the piece in a cold 10-percent sulphuric acid bath until all traces of flux have been removed, then wash the piece in fresh, running water to remove the acid.

BUAER TESTS FOR NONFERROUS METAL

As in the case of ferrous metals, the Bureau of Aeronautics requires an aircraft welder to pass certain qualification tests before he is permitted to weld aluminum.

Three types of welds with aluminum sheet are required. These are the flanged joint on metal 0.083 inch in thickness, shown as (A) in figure 34; the butt joint on metal from 0.083 inch to more than 0.125 inch in thickness, shown in (B) and (C) ; and the fillet T-joint, illustrated in

(D) of figure 34.

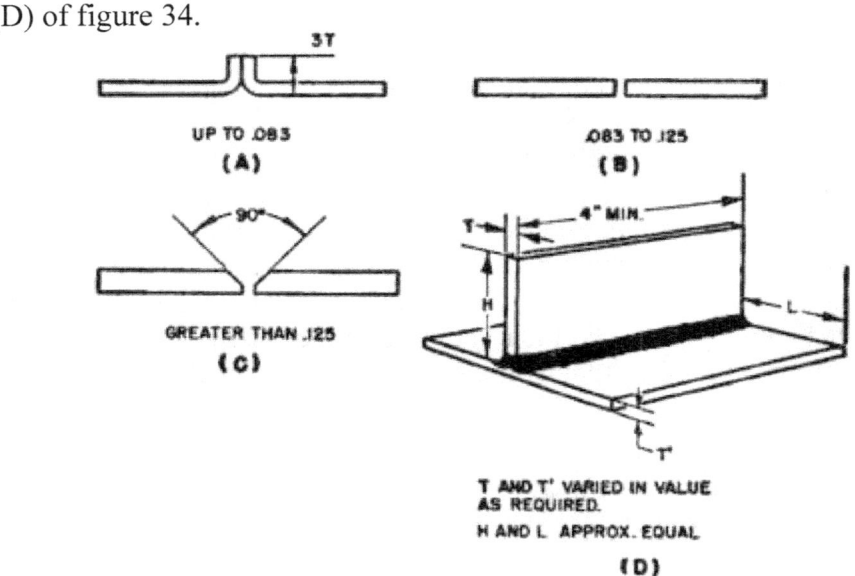

UP TO .083
(A)

.083 TO .125
(B)

90°

GREATER THAN .125
(C)

4" MIN.

T

H

L

T'

T AND T' VARIED IN VALUE
AS REQUIRED.
H AND L APPROX. EQUAL
(D)

Figure 34.—BuAer qualification tests for aluminum welding.

The flanged joint is made on aluminum sheet pieces approximately 6 inches square. The height of the flanges turned up should be two or three times the thickness of the sheets. Clamp the flanges together at first, then tack-weld the edges to preserve the alinement. The flanges are then melted down, with welding rod and flux being added as required. The finished weld should show a smooth, well-rounded bead on both sides of the joint with little or no evidence to indicate that the sheets had been flanged.

Thicknesses of metal between 0.083 and 0.125 inch need not be beveled or flanged. In making the butt joint in sheet aluminum of more than 0.125 inch in thickness, bevel the edges at 45° angles to form an included angle of approximately 90°.

A T-pillet weld, approximately 4 inches in length, is another BuAer test. It is well to remember that the inspector may exert his prerogative and specify the most unfavorable combinations of thicknesses in the two plates.

To inspect these joints the inspector cross-sections each and checks for unevenness, undue roughness, cracks, blowholes, la«k of penetration, lack of fusion, and so forth.

The flange joint of aluminum on a non-heat-treatable alloy must withstand a bend test of 180°—that is, until the metal is completely bent back on itself along the line of the welded joint—without showing cracks in either the base or weld metal.

ALUMINUM CASTINGS

Aluminum castings are of complicated design, and the thickness of aluminum in their composition will vary from section to section. This means that aluminum castings are considerably more susceptible to heat strains and cracking than is aluminum sheet, and thus many castings are heat treated for added strength. Welding such heat-treated castings tends to destroy the effect of heat treatment, and unless facilities are available for reheat-treating after welding, they should not be welded.

A broken aluminum casting destined for welding should be cleaned with a wire brush and carbon tetrachloride to remove all traces of oil, grease, and dirt.

Edges of sections heavier than three-sixteenths inch should be beveled at about 45° by mechanical means. Such conditions can be done on thinner sections with the torch and puddling rod, the latter being a tool used to break up the oxide in aluminum weld metal when no flux is used. Figure 35 shows three puddling rods with various shaped ends. These instruments are made from pieces of ^4-inch steel-welding rod about 12 inches in length, one end of which is heated with a torch and flattened to a width of about three-eighths inch, then ground smooth. The other end is bent over to form a handle.

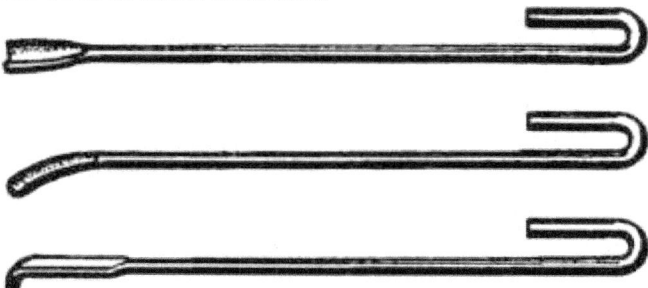

Figure 35.—Puddling rods for aluminum welding.

The procedure for beveling a V when repairing a cracked casting is to heat a 2-inch section of the crack by brushing the flame back and forth, holding the flame several inches from the metal. The metal will rapidly become hot enough for the flattened end of the puddling rod to scrape out a V reaching to the bottom of the seam. Keep the puddling rod out of the flame.

Make certain that all unsoundness around a crack in a broken casting is melted or cut away before beginning the weld. Aline the broken part, then use light iron bars and clamps to hold it in position for preheating and welding.

Fasten the clamps snugly but not so tight as to strain any part of the casting, especially in thin sections. Aluminum, when hot, is very weak and will crumble under what would ordinarily be only slight strains.

If the casting is large or has intricate sections, preheat it slowly and evenly in a furnace. If small, or if the break is near the edge and in a thin section, preheat the area immediately around the break with the torch flame. In any case, remember that cast aluminum must be heated slowly to prevent cracks in the part of the casting near the flame.

Preparation and preheating completed, tack-weld the edges of the crack. Begin welding in the middle of the seam and weld toward the ends.

As in sheet, cast aluminum melts at comparatively low temperatures, and the heat of the molten weld pool is not great enough to melt the filler rod. As the weld progresses, hold the rod barely out of the inner flame in order to melt it. For ordinary castings, a 43S rod is used.

Flux is necessary in welding aluminum castings to float the aluminum oxide to the surface. The tough oxide film can be broken up by the action of the puddling rod, but unless flux is present, the oxide skin broken up by puddling may remain in the weld. Oxides cause poor fusion.

When the weld is completed, scrape off the excess molten metal with the puddling rod, then allow the casting to cool slowly before washing off the flux.

Holes in castings are welded similarly to cracks. Before starting the repair of a hole, melt or cut away its sides to eliminate any pockets in the weld and permit easy access to the hole.

MAGNESIUM

Magnesium, lightest of the structural metals, is currently obtained by a war-born process which "mines" magnesium salts from the sea. As a matter of fact, magnesium has been produced commercially since 1916, using brines from wells. During World War II several plants were constructed to mine magnesium from the waters of the Gulf of Mexico.

Magnesium alloys are used in aircraft construction because of their light weight (magnesium weighs two-thirds as much as aluminum), their strength, and their excellent machinability. Pure magnesium, like pure aluminum, is not very strong, but by adding other metals—particularly aluminum, zinc, tin, manganese, or combinations of these metals—alloys have been developed which possess great strength for their light weight

Repairing magnesium alloy parts by welding is not permitted if the metal is used in a structural member. Magnesium alloys used in structural parts are heat treated, and such alloys are not weldable for the same reason that heat-treated aluminum alloys are not weldable—because it is impossible to furnish the welded section with the required strength.

A weld in sheet or plate magnesium alloy 0.05 inch or thicker is from 60 to 90 percent as strong as the base metal, depending upon the type of alloy, its thickness, and its temper. Failures in the weld do not as a rule occur in the weld zone, but in adjacent base metal which has been melted and resolidified.

Those magnesium alloys which are easily welded and which may be welded to each other are AM3S, AM52S, and AM-C52S, produced by the American Magnesium Co.; and the corresponding Dowmetal alloys M. FS, and Fa-1, manufactured by the Dow Chemical Co. Two other magnesium alloys used in aircraft—AM-57S or Dowmetal J and AM-C57S or Dowmetal J-1—can be welded only in relatively short seams. Gas welding of other wrought aluminum alloys, while possible to a certain extent, is not generally recommended.

Magnesium alloys cannot be welded to aluminum alloys or to other metals. Welding repairs on magnesium alloy castings are not generally recommended, although cast or forged fittings may be welded into magnesium sheet structures.

The filler rod for magnesium alloy welding should be of

the same composition as the base metal. Welding rod for magnesium welding comes in the alloys and sizes indicated in table 6.

Table 6.—WELDING ROD FOR MAGNESIUM WELDING

The approximate sizes of welding rods which should be used with the various thicknesses of magnesium sections are shown in table 7.

Table 7.—SIZES OF WELDING ROD FOR MAGNESIUM WELDING

The filler rod comes covered with a dichromate coating which must be removed with steel wool or a wire brush before using the rod in welding. If a welding rod is not available, use a strip cut from the magnesium to be welded. Remove all traces of flux from any unused portion of welding rod immediately after finishing a weld; otherwise it cannot be used again.

Select a welding tip whose diameter in relation to the metal thickness checks with the following table:

TobU •.—WILDING TIP SIZES

The oxyacetylene welding flame protects the molten magnesium from oxidation because of the nonoxidizing properties of the burned gas with which the weld area is surrounded.

The reason for keeping oxygen from molten magnesium alloys will be readily apparent when it is remembered that powdered magnesium is used for flares. While welding, bathe the weld area with the outer envelope of the torch flame to prevent air from reaching the molten metal. Make certain that the torch flame is adjusted to neutral so that all of the oxygen coming through the torch is consumed by the burning acetylene gas. Ordinarily, little trouble with oxidation is encountered if the two precautions just mentioned are carefully observed.

Slant the flame at an angle of 30° to 45° to the surface of the work. To avoid burning the metal, the angle for thin sheet metal should not be more than 30°.

The flux for welding magnesium is a special mixture. Make a paste of the powdered flux by adding one part water to two parts flux, mixing a fresh supply daily.

In magnesium welding, the flux must be completely removed after welding. Consequently, only butt welds can be made, as any other type of joint affords too great an opportunity for the flu.i to hide away in pockets and corners where it cannot be removed.

Lap welds and flange welds in which the flange is not melted down are thus out of the question in welding magnesium alloys. Fillet welds—welds made in a corner— should be made only in tubing. When making fillet welds in tubing, drill holes near the welds to permit flux removal by means of rinsing.

Castings or extrusions can be welded to sheet parts if the portion of the casting or extrusion to be joined to the sheet part is of approximately the same thickness as the sheet. Sections of slightly different thickness may, at times, be welded to each other, if the heavier section is preheated so that both edges of the seam will begin melting at the same time.

The method of preparing a butt joint depends on the thickness of the sheet metal. Sheet magnesium alloy up to 0.040 inch in thickness should be flanged up about three thirty-seconds inch to the angle as indicated in figure 36, and melted down into a butt joint.

Butt joints on metal from 0.040 to 0.125 inch in thickness are neither flanged nor beveled, and a space of one-sixteenth inch is allowed between the edges of the joint. For heavier gages, use a cold chisel to make notches along the edges, one-sixteenth inch deep and three-sixteenths inch apart, to aid penetration. For butt joints in metal thicker than 0.125 inch, bevel down each edge 45° for about two-thirds of the sheet's thickness to make a 90° included angle for the V. Do not bevel magnesium alloys by using a torch.

The space between the edges should be one-sixteenth inch for metal from 0.125 to 0.250 inch in thickness. On greater thicknesses, leave a %-inch space between the edges and build up the weld bead in more than one pass, laying down a deposit of weld metal each time. In this case, ordinary precautions to avoid oxidation of metal are not enough, and the weld area should be completely enveloped in an atmosphere of carbon dioxide.

Remove oil or grease from the weld area with carbon tetrachloride, or a hot alkaline cleaner, then use a file, wire brush,

GAGES UP TO J040   O40T0.I25

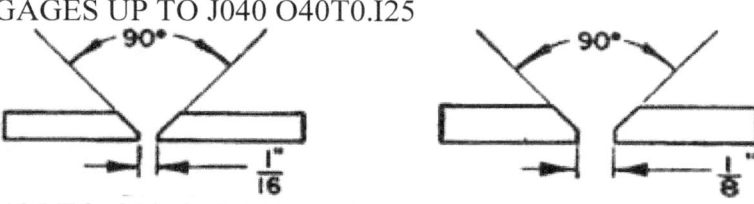

.125 TO .250   .250 AND UP

Figure 36.—Preparation for welding of butt joinfs in sheet magnesium.

or abrasive cloth to clean and brighten the edges of the joint, including the surface of the metal three-fourths inch back from the edges. Then brush the flux paste on the underside of the parts, also coating the welding rod.

If the job is repairing a closed structure, and the underside of the seam is inaccessible for application of flux, the metal will oxidize unless the part is filled with carbon dioxide while making the weld.

Making a Magnesium Weld

Start the weld by tacking the seam at intervals of 14 to iy 2 inches, depending upon the type of work. If the sheet warps while tack-welding, it can be straightened with a wooden hammer.

After tack-welding, coat the tacks with flux and then weld completely a piece of the seam from 1 to 2 inches long at each end. This reduces the tendency of the metal to warp while the weld is being made. Shrink cracks at the end of the seam can be eliminated by adding an excess of weld metal at that point. Use as little heat as possible.

Your success in welding, except in the previously mentioned case of thick metal, depends upon being fast and efficient with the torch. The job should be accomplished in one pass. Going back over the seam causes cracks and distortion.

Each of the two major producers of magnesium in this country recommends a different way to weld magnesium. One prefers keeping filler rod in the molten pool of metal, which means getting enough flux on the filler rod before starting so that the rod will not be taken out of the puddle to renew the flux. The other producer advises using the same torch and rod technique for magnesium as used for aluminum. The method with which you attain the greatest success should be used.

If any part of the weld begins to oxidize, carefully scrape out the oxidized portion of the weld before continuing the operation.

Buckling which you are unable to avoid through accurate lining up of the parts and closely spaced tacking may be eliminated with a wooden hammer while the metal is still warm.

Let the weld cool slowly. When the weld is cool enough to handle, scrub the accessible parts lightly to remove most of the flux. Then put the part in hot water (160° to 200° F.) and soak off the large particles of flux adhering to any inner parts of the weld which scrubbing does not affect. Keep fresh water flowing constantly into the tank where the part is soaking to prevent accumulation of dissolved flux.

This done take the part out of the hot water and soak it for 10 minutes in a 1 percent citric-acid solution. Gas bubbling from the vicinity of the weld will loosen the flux particles. If the part is so made that gas cannot escape readily, rock or turn it frequently to prevent the formation of gas pockets where the flux may not be completely removed.

When removing the part from the citric-acid solution, drain it thoroughly and rinse it carefully in clean, cold water. This rinsing must be thorough in order to flush away the loose pieces of flux remaining inside the part.

As an alternative to the citric-acid bath, submerge the part from 30 seconds to 2 minutes in a chrome-pickle solution. A chrome-pickle dip consists of a water solution containing iy 2 pounds of sodium or potassium dichromate, 0.2 pound of magnesium sulfate, iy 2 pints of concentrated nitric acid, and 1 gallon of water at room temperature. Follow this dip with a thorough rinsing of the part in cold running water and then in hot water to aid the drying process.

It is important that the part be dried immediately after the rinsing operation. A dip in clean, boiling water will further this process.

The joints which the Bureau of Aeronautics require for qualification on magnesium alloys are practically identical with those required for welding aluminum. A study of figure 36 furnishes a review of these requirements.

INCONEL

Inconel is the trade name for a nickel alloy metal composed of approximately 75 percent nickel, 12 to 15 percent chromium, and 9 percent iron, with small percentages of carbon, copper, manganese, and silicon.

Wrought Inconel is made up in bars, plates, sheets, strips, i wires, and seamless or welded tubing. It is very resistant to corrosion and also to fatigue brought on by high temperature. Fatigue means that the metal may "get tired" and fail under vibration or shock stresses. Because of its corrosion and fatigue-resistant qualities, Inconel is sometimes used in aircraft engine exhaust systems.

Inconel Welding

To weld Inconel, a flux is required on both the filler rod and the top and bottom of the joint. If a commercial flux is not available, satisfactory paste may be made by mixing three parts of sodium fluoride and one part of an equal mixture of borax and boric acid.

The filler rod should be of the same composition as the base metal. Select a rod of correct size for the thickness of the metal to be worked on—that is, a rod one-sixteenth inch in diameter for metal one-sixteenth inch thick, and so forth.

Welding Inconel requires a carburizing flame (one with a slight excess of acetylene) rather than a neutral flame. Adjust the flame until the feathery middle cone is about iy 2 times as long as the brilliant inner cone. Hold the torch flame so that this feathery edge just touches the surface of the metal.

The welding tip should be the same size or one size larger than that used in welding steel of the same thickness.

When working on light gage Inconel, use the forehand method of welding, and tip the torch to an angle of about 46° to the work.

The molten metal must not be puddled, and the rod should be added to the weld with a minimum of disturbance. Puddling and any unnecessary disturbance of the molten pool burns out the deoxidizing element (Silicon) of the metal and results in brittle and porous welds. Add rod to the weld by allowing it to melt in the molten base metal.

Inconel Joints

Butt welds can be made on sheet Inconel for either a close (rigid) joint or an open joint.

In welding an open-type butt joint, avoid heat-distortion of the metal by spacing the edges to be joined at a slight taper. Start welding from the ends which are closer together and the edges will join as the weld proceeds until the gap disappears, Use this type of taper spacing on thin sheets up to 0.0625 inch in thickness. Start at the closed end with a space equal in width to the thickness of the metal and widen it toward the other end at the rate of 14 mcn for each foot of seam length.

A weld, once begun, should be completed without interruption. Should interruption occur, however, the weld must be reheated to a bright red color for about an inch back of where the previous weld left off.

238500° 5H 8

If a closed or rigid type of butt joint is to be made, set the edges up parallel, leaving a space between them equal to the metal thickness. It may sometimes be necessary to trim the edges of the crack until a space of the proper width remains. Then the seam is tack-welded at points 2 inches apart, and a small amount of flux is applied to each tack. Complete the weld, beginning at the end opposite the first tack.

Inconel tubing is handled in the same manner as sheet for welding butt joints. Set up the parts with the end separated a distance equal to the thickness of the tube wall, then tack weld. About four tacks spaced evenly around the circumference of the tube will suffice for tubes under 2 inches in diameter. For larger tubes, use more frequent tacks.

Full penetration is fully as important in Inconel welding as in any other metal or with any other type of joint. A butt weld on sheet or tubing should show a slight bead, or thin ridge of weld metal, on the under side.

In addition, build up such a butt weld so that it extends above the surface of the parts joined (this is called top reinforcement) a distance about equal to the base metal thickness. Such top reinforcement should also extend out on each side sufficiently far beyond the edges of the seam to make certain that good fusion is obtained on both edges.

Fillet welds can be made successfully on Inconel. Gruard against cracking or pulling out, as Inconel is weak and brittle when after having been melted, it begins to harden.

When fillet welding Inconel fittings near the end of a piece of sheet or tubing, one method of overcoming this weakness of Inconel is to heat the metal to a bright red completely to the end before beginning the weld. When the weld is finished, the part should be reheated to reduce the stresses and prevent buckling and cracking.

As in welding other metals, use jigs wherever possible to hold the pieces in line.

A copper backing plate should be used to hold the molten metal at the root of the weld. This plate should be grooved

to a depth of one thirty-second inch to allow full penetration so that a slight bead will be present on the under side of the finished weld.

Adequate ventilation is essential when welding Inconel because the fumes from the melting flux are poisonous.

MONEL METAL

Monel metal is a trade name for another nickel allot. Monel contains 60 to 70 percent nickel, 23 to 30 percent copper, plus small amounts of carbon, iron, manganese, and silicon.

Monel metal is manufactured in wrought bar, sheet, tube, and wire forms as well as in castings for special purposes. It is strong, tough, ductile, and highly resistant to corrosion from many acids and alkalies.

Practically all welding processes can be used on Monel. A flux is necessary to protect the hot metal from the oxygen in the air, and also to dissolve any oxides present in the metal.

The filler rod must have the same composition as the Monel. If a conventional welding rod of the right type is not available, a strip cut from the sheet of Monel will do.

Hold the torch at a 45° angle to the surface of the work and use the backhand method so that the flame can envelop the molten weld metal and exclude oxygen.

Sheet Monel metal stock should be welded in a jig in order to keep the edges alined and reduce buckling.

When a butt joint is required in welding thin Monel metal sheet up to 0.0625 inch in thickness, it will be necessary to turn up flanges about one-sixteenth inch in height at an 80° angle. The edges of the flanges will thus come together at the top at a 20° angle, and the flanges themselves will not be quite parallel—an antiwarping trick. Before welding such flanged butt joints of Monel, apply a thin flux to the flanges and tack-weld the seam at an even height.

The flanges must be melted down as rapidly as possible to avoid burning the metal.

ToW. 9 —MELTING POINTS Of AIICtAFT METALS

Metal

Melting point (°F.)

Aluminum, cast, 8 percent copi>er_
Aluminum, pure -
Aluminum, 5 percent silicon
Bismuth
'"Brass, commercial high
Bronze, tobin
Bronze, manganese
Chromium .
Copper, deoxidized
Copper, electrolytic..
Iron, cast
Iron, malleable
Iron, wrought
Lead
Molybdenum. Monel metal _ Nickel
Silver
Steel, hard (0.40-0.70 percent carbon)
Steel, low carbon (less than 0.15 percent carbon)
Steel, medium (0.15-0.40 percent carbon)._.
Steel, manganese
Steel, nickel 3^4 percent
Steel, cast r
Stainless steel (18 percent chromium, 8 percent nickel).
Stainless steel (18-8 low carbon)
Tin
Vanadium
Zinc, cast or rolled
1, 175 1, 220 1, 117 520 1, 660 1, 598
1, 598
2, 740 1, 981
1, 981
2, 300 2, 300 2. 900
620 4. 532 2, 400 2, 646
1, 762
2. 500 2, 700 2. 600 2. 450 2. 600 2, 600 2, 650
2, 640 450
3, 182 786

Butt joints on sheets heavier than one-sixteenth inch should be beveled. The edges should be separated about one thirty-second inch at the end where the weld is to begin, and the space should gradually widen about three-eighths inch for each foot of seam length.

Welding thick pieces requires that the bead be built up to its full height as the weld progresses. Welds built up in layers are weak because each layer will be separated by an oxide coating. The weld must also be built up well above the surface of the base metal. This creates a margin of weld metal which may contain any oxide film and impurities which have been floated to the surface. Such impurities will be on top and can be ground or machined off, leaving a sound, clean weld.

To attain a certified welder's rating, tests on butt and T-fillet welds in both thin and heavy Monel and Inconel are required. The inspector may require several aircraft parts made of these alloys to be welded.

The joints will be inspected for penetration and fusion, and subjected to a bend test in the same manner as were those on stainless steel.

MELTING POINTS

One of the most important characteristics of metals from a welder's standpoint is the melting points of the various aircraft metals. This knowledge is applied, for example, in deciding whether to weld or braze two dissimilar metals.

If there is a wide range between the melting points of two metals, proper fusion would be extremely difficult to obtain. If a brazed joint would serve the purpose, it would be preferable in such case, because in brazing, the two metals do not require melting.

Table 0 lists the melting points of the most important aircraft metals.

QUIZ

1. What is meant by nonferrous metal?
2. Why is an aircraft welder not permitted to weld structural parts made of heat-treated aluminum alloys?
3. What size tip should be used in welding aluminum?
4. Why must flux be used in welding aluminum?
5. To what welding principles should special attention be given when welding aluminum?
6. What joints are required by BuAer tests for aluminum sheet?
7. What facilities must be available when aluminum castings are welded?
8. When is the welding of magnesium alloy parts forbidden f
9. Can magnesium alloys be welded to other alloys or metals?
10. How should the torch flame be adjusted when welding magnesium alloys? Why is this important?
11. What types of weld joints may be made on magnesium alloys?
12. WhatisInconel?
13. What type of flame is used for welding Inconelf
14. Why is ventilation important when welding Inconel?
15. What is the use of the copper backing plate when welding Inconel?
16. What two types of weld are best adapted for use on Inconel ?
17. What is Monel metal ?
18. What welding processes can be used on Monel metal ?
19. Assuming that either a brazed joint or a welded joint would serve the purpose, which type of joining should be used for two dissimilar metals having widely different melting points?

CHAPTER 6
WELDING AIRCRAFT TUBING

TUBE WELDING TECHNIQUES

Iii repairing aircraft parts by welding, the structural mechanic will more often be concerned with tubing than with flat stock, since engine mounts, landing gears, and collector rings, which are the parts usually repaired by welding, are made of steel tubing. Because the shapes of the tube ends involved in welded joints and splices are not as simple as the parts of a joint in sheet or plate, preparation of tubing joints requires more time and more careful workmanship.

The welding of tubing involves the fundamental techniques used in the welding of flat stock, but several practices may be required on the same joint. For example, a butt joint in tubing might be welded partly by forehand and partly by backhand welding. This same joint, if welded in the fixed position, might require welding in the flat, vertical, and overhead positions.

In addition to adapting his techniques to tubular shapes, the aircraft welder must have an understanding of the reaction of high-strength alloys under the welding flame if he is to successfully weld tubing made of these alloys.

TUBING MATERIALS

At one time, fuselage and wing structures were made of mild carbon steel (S. A. E. 1025) almost exclusively. This steel has been replaced for most uses by chrome molybdenum steel (S. A. E. X4130) because of the latter's superior tensile and fatigue strengths, its corrosion resistance, and its shock impact resistance. For these reasons the characteristics of this metal under the welding flame must be clearly understood.

Stainless steel of the 18-8 type is also widely used in the fabrication of collector rings and other exhaust system parts because of its great corrosion resistance and its high tensile strength and ductility.

TUBING JOINTS

Tubing joints are different from joints in flat stock—not only in shape, but in the amount of strength developed. For example, the plain butt joint in sheet or plate is seldom used in tubular construction. As shown in figure 37, welding such a joint develops annealed areas or zones parallel to the bead that are so weakened as to fail by buckling or wrinkling. Consequently, for butt joins a weld called the fishmouth joint or a weld called the scarf joint is used.

A FiSHMOUTH joint is a tubular joint used in joining two pieces of tubing end to end, in which the edges are cut to resemble a fish's mouth. For pieces of equal diameter a BUTT joint with the joining edges of both pieces cut in matching fishmouths is used. For pieces of unequal diameter, a reduction joint with only the end of the larger piece fish-mouth-cut is used.

A scarf joint is a joint between two members in line with each other, in which the joining ends of one or both pieces are cut diagonally at an angle of about 30° from a center line. This is called a scarf cut. In welding aircraft tubing, scarf joints are used both as butt joints and reduction joints.

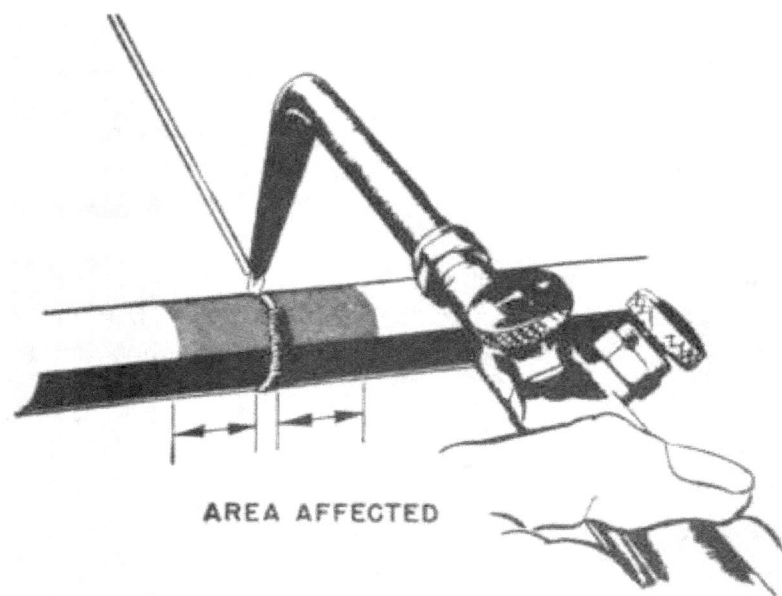

AREA AFFECTED

Figure 37.—Annealed area* *et up in butt-welded joint.

The joint referred to above as a reduction joint is that made between two members of unequal diameter or width, both members being of the same general plane—that is, not at an angle to each other. Reduction joints are used in the welding of aircraft tubing to join tubes end to end for greater length as in longeron construction; to repair defective sections of tubing, or to brace a section of a piece of tubing. When additional length is the main purpose, the end of the smaller tube is telescoped into the end of the larger tube, far enough for adequate bracing. A welded joint of this type is often called a telescope joint. When repair or bracing of a central section is the main purpose, a short section of larger tubing is slipped over the smaller tube like a sleeve.

Where greater strength than that obtainable with either the scarf or fishmouth butt joint is desired, splices using sleeves are commonly used. These splices are particularly adaptable to the repair of tubular structures, such as engine mounts. Specifications for repairs using these splices will vary with individual aircraft. One rather common restriction is that no welds can be made in the middle third of a tube section, as that is the section most likely to be distorted by compression or bending stresses.

CHROME-MOLYBDENUM WELDING TECHNIQUES

Chrome-molybdenum is easy to weld, has a high initial strength, and is extremely shock-resistant. Code name for the most common kind of chome-moly is S. A. E. 4130.

The oxyacetylene flame is generally preferred for .thin-walled tubing made of chrome molybdenum. For material 0.093 inch and thicker, the electric arc is preferred, as the heat zone is narrower, resulting in less distortion from expansion.

The welding technique with the oxyacetylene flame is the same as that required for the carbon steels, except that the area surrounding the weld should be heated to 300°-400° F. before starting the weld. This is necessary, as a sudden application of the flame without some preliminary heating sometimes causes cracks in the heated area.

The flame should be directed on the metal during welding at such an angle that preheating takes place ahead of the weld. A soft neutral flame must always be used, as an oxidizing flame burns and weakens the steel. A carburizing flame renders the metal brittle and may also cause cracks on cooling. The volume of flame should be just large enough to reduce the base metal to a melting state so that the proper fusion will take place. Overheating will set up severe stresses, and will cause excessive grain growth, which contributes to low strength in the welds and in the adjacent area of the base metal. The weld metal should be protected from the air as much as possible while hot.

Chrome molybdenum tubing is quite generally welded with a low carbon (mild steel) rod, welds of satisfactory strength and ductility resulting if the correct techniques are used.

A chrome molybdenum rod may be used for joints requiring high strength, if the part can be heat treated after welding. No flux is required. Better results will be obtained if the welds are not begun at the raw edge (as of a gusset), since the metal may crack. Nor should the torch be lifted suddenly from the metal at the end of a weld, as this may cause a pinhole, at which point stresses would be concentrated.

The AN Repair Manual (AN-01-1A-1) recommends the tip sizes indicated in table 10 for oxyacetylene welding of chrome molybdenum tubing.

Table 10.—TIP SIZES FOR OXYACETYLENE WELDING OF CHROME MOLYBDENUM TUBING

Welds in aircraft tubing must not be dressed (filed or smoothed down) unless further welding is to be performed on the dressed section. If the weld is on thin tubing—less than 0.040 inch—it should not be wider than one-quarter inch.

Precautions

Before beginning a weld, clean the parts thoroughly by taking off all grease, paint, or other foreign substances from the section to be repaired.

Remember also that chrome-molybdenum steel is strong and tough when cool, but it is weak at high temperatures. Because of this trait, the metal may crack if subjected to even h comparatively slight stress while in a white-hot condition. Therefore, the rule to follow with hot chrome-moly is— handle with care.

Such handling requires that the following two specific precautions be taken. (1) Never use welding jigs that will hamper expansion or contraction of welded members. These two familiar reactions—contraction and expansion—which occur to some degree in all metals when heat is applied, are present when welding chrome-molybdenum aircraft tubing. (2) Be extremely careful to avoid overheating and burning the metal when welding near or at an edge. Edges heat up quickly, and welds started at a raw edge may cause the metal to crack.

It is also important to prevent loss of metal thickness from excessive scaling (formation of iron oxide).

The torch should not be lifted suddenly from the metal at the end of a weld because of the danger of creating a pinhole. A good technique is to "fan" the end of the weld with flame when finishing the bead. This prevents sudden chilling of the puddle.

Cracking may be prevented by cutting down the strain caused either by the weight of the parts or by the restriction of normal expansion and contraction. For this reason, when welding chrome-molybdenum tube assemblies, good technique demands that one end of the web member be completely welded to the flange member of a truss, and the weld be allowed to cool before welding of the opposite end of the web member is begun.

Similarly, joints of an assembly in which several members terminate should be welded first and allowed to cool before attempting to weld the opposite ends. Assuming that these members are connected to similar joints at their opposite ends, heavy clamps, chill plates, or wet asbestos should be used on the members near the weld. This procedure aids in preventing expansion, caused by heat travel, to adjacent parts. As can be readily seen, welding a number of joints terminating in such an assembly naturally takes more time and requires more heat than merely welding a single member into a joint. Consequently, there is a great amount of expansion.

In welding fittings, shrinkage strains and the consequent danger of cracking can be checkmated by starting the weld at the fixed end of the seam and working toward the open end.

It is good practice to relieve the stresses of alloy steel parts after welding by a process known as drawing off. This means heating the entire part uniformly to a temperature between 1,150° and 1,200° F., then permitting it to cool slowly. If equipment is not available for heating the entire piece, the same effect can be obtained by using the torch to .heat the part for a radius of several inches around the weld area.

If a joint is to be secured both by riveting and welding, aline the rivet holes and complete all welding before the rivets are driven. If this procedure is reversed, expansion and contraction forces develop a shearing stress on the rivets and tend to stretch, or elongate, the rivet holes.

In welding tubing assemblies, warping can" be controlled by the use of sufficient clamps and the proper kind of jigs.

Quarter welding, a variation of skip welding, is used to help prevent distortion. To quarter weld, tack-weld the tubing joint at four equidistant points, then weld diametrically opposite segments from tack to tack. Skip welding throughout the piece also aids in reducing distortion. A typical tubing repair in which quarter welding is recommended will be discussed later in this chapter.

MAJOR REPAIRS

Unless the damage to members of a steel tube structure is comparatively slight, the following procedure should be followed: remove the injured part and weld in either a partial replacement tube or an entirely new piece of tubing. The tubing used for telescope reinforcement or splicing must be of at least the same tensile strength and wall thickness as the original member.

Tubes and sections are always cut out with a hacksaw— never by oxyacetylene flame cutting. Tubes inserted as replacements are then joined at their ends by means of a splice.

Splicing of partial tube replacements may be done by using a replacement tube of the same diameter plus either internal or external reinforcing sleeves, or by using an external replacement tube of the next larger diameter. Such a replacement tube is spliced to the stub ends of the original tubing.

Each type of splice has its particular advantage and use, even though the methods involved in making them are essentially the same. If the original damaged tube included castings or fittings that have been made to fit the tube, then the spliced replacement tube must be of the same diameter as the original tubing. Therefore, either internal or external reinforcing sleeves under or over the splices can be used. If no fittings are attached to the original tubing, the replacement can usually be an external tube.

Rosettes are often added to a splice to increase the shear strength of the repaired member. Holes are drilled in a staggered formation around the outer tube. These holes should have a diameter equal to one-fourth of the tube diameter, but not less than one-fourth inch. The rosette is welded around the inner edge of the hole into the smaller telescoped tube after all other welds on the joint have been completed.

Only one partial replacement tube can be inserted in any one section of a structural member. If more than one tube

DRILL HOLE IN TUBING. WELD AROUND CIRCUMFERENCE TO INNER SLEEVE.

in a joint is damaged, remove the entire joint and insert a new, preassembled and welded joint of the proper design.

If a web member is damaged at a joint so that it is impossible to retain at that point a stub long enough to permit splicing on replacement tubing, put in an entirely new web member.

If a continuous longeron is damaged at a joint, make certain that the replacement tube splices on either side of the joint are far enough from the joint to prevent weakening the weld. First cut loose the web member at the affected joint and remove the damaged section of tubing. Then splice the replacement tubing to the stub ends of the original section of longerons. Finally, weld the web member to the new section of longeron tubing. Use wooden braces to hold the tubes in alinement during the repair.

INNER AND OUTER SLEEVE REINFORCEMENTS

In figure 39 is shown a partial replacement tube spliced to the original tubing by means of inner reinforcing sleeves— a recommended Navy practice. In this procedure, a minimum of welding is required, and consequently there is less chance of weakening or distorting the tubing. This method has the advantage of giving a smooth outer surface to the repaired section.

In making this repair, first make diagonal cuts in the affected tubing, ascertaining that the cuts are located away from the middle third of the affected tube section. When the part has been removed, file off the burr, or roughness, from the edges of the cuts, select a replacement tube matching the damaged original tubing in both diameter and wall thickness, and cut from it a length one-fourth inch less than that of the removed section, by means of similar diagonal cuts on the ends. This will leave an %-inch gap between each end of the replacement tubing and the original tubing.

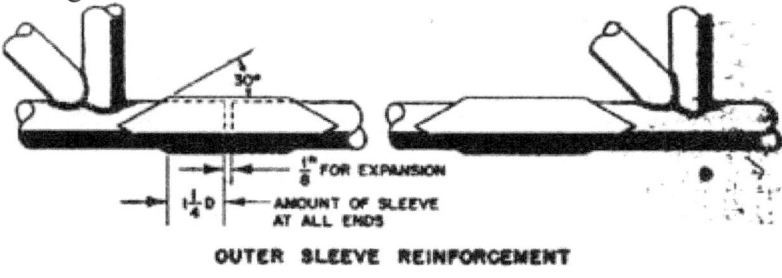

OUTER SLEEVE REINFORCEMENT

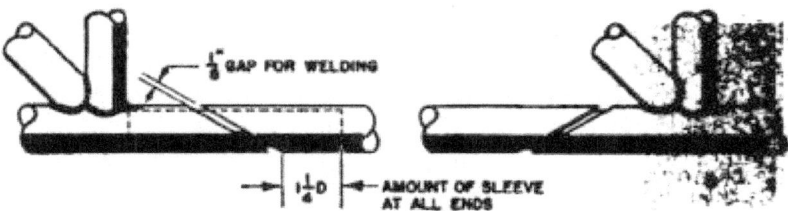

INNER SLEEVE REINFORCEMENT Figure 39.—Inner and outer slwv* spile*.

Now cut two reinforcing sleeves, sawing straight across (not diagonally) the tubing. For this purpose, select tubing of the same wall thickness as the original, and with an outside diameter equal to the inside diameter of the original tubing. The sleeves should make a snug fit inside the original tubing—clearance between sleeve and tubing should be no more than one sixty-fourth inch. These inner sleeves should be long enough so that either end of a sleeve is not
us
less than 1*4 tube diameters from the original cuts in the original tubing and the replacement tube.

The success of the splicing process depends on following a logical step-by-step procedure, as set forth in the following pages.

Set up a brace arrangement to support the structure while welding. Figure 40 illustrates how a brace replaces the damaged tubing in holding the vertical members in aline-ment while working.

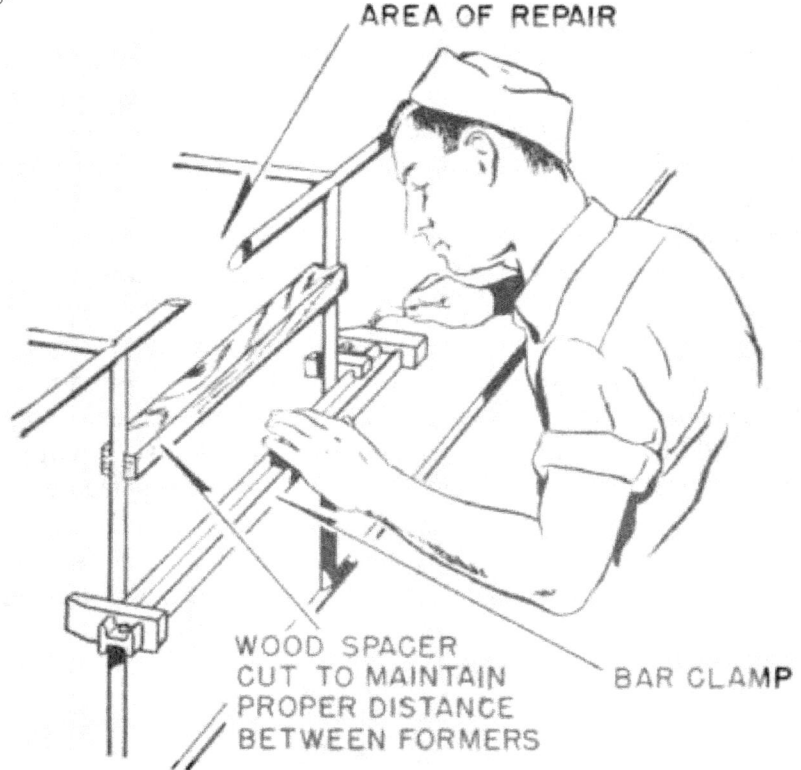

AREA OF REPAIR

WOOD SPACER CUT TO MAINTAIN PROPER DISTANCE BETWEEN FORMERS

BAR CLAMP

Figure 40.—Brace for alining members.

Dip the replacement tube and inner sleeves into hot (about 165° F.) Paralketone (raw linseed oil), to help prevent corrosion. Wipe the Paralketone from the outside of the replacement tubing and sleeves.

288600*—S3- —9

Make a small mark on the outside of each original tube stub end, halfway along the diagonal cut, as shown in figure 41. Then measure off a distance 2*4 diameters long from the nearest end of each diagonal cut on the original tubing. Center-punch the tube at these points and drill No. 40 holes— with the drill held at a 90° angle to the surface of the tubing. When the hole is started so that the drill will not jump out. slant the drill toward the cut and continue at a 30° angle. Remove burrs from the edges of the holes with a ^4-inch drill

Take a length of ^-inch welding or brazing wire, thrust one end through the hole just drilled, and push it out the diagonally cut open end of the original tubing. Repeat this at the other stub end. Use these wires to draw the sleeves into the tubing.

Next, weld the end of each wire protruding from the open end of the tubing to the inside of one of the inner sleeves, as may be seen in figure 41. To help draw the sleeve into the tube, bevel the ends of the sleeves to which the wires are welded.

Make a narrow mark around the center of the reinforcing sleeves.

IAS UNCI* jf

/ DRILL N0.40 HOLE INTERNAL * SLANTED AND POSH imciuwM.

£ WCLOINfl WO THRU BEVEL ENOSTcTaIO IN SUO-
■• INS MARK CENTS). WELD

MARK CENTER. WELD ROO EXTEN04N* PROM END Of TUBING TO INSIDE ENO OP SLEEVE.

INSERT INTERNAL SLEEVE INTO SPLICE MEMBER. AUNE.THEN PULL WIRE TO CENTER. COMPLETE WELD THEN TRIM WELD OVER HOLE.

41 •••^How drcNAf lnn#r

At this juncture, push the inner sleeves into the replacement tube so that the point where the wire is welded to the sleeve is 180° from the drilled hole. If the drilled hole is at the bottom of the tubing, the inner sleeves are placed so that the point at which the wire is welded is at the top. If the inner sleeve fits too tightly in the replacement tube, chill the sleeve with dry ice or in cold water. If the sleeve persists in sticking, polish it with emery cloth.

Aline the stub ends of the original tube with the replacement tube.

Now begin pulling the end of the wire protruding from the drilled hole. Pull the sleeve along until the center mark on the sleeve is directly in line with the center mark on the diagonal cut. When these marks line up, the sleeve is centered beneath the joint, as shown in figure 41. Repeat this process for the other sleeve at the opposite end of the replacement tube.

Bend the pulling wire over the edge of the hole so as to hold the sleeve in position, and weld the inner sleeve to the original tube stub and replacement tube at one end. This fills the %-inch gap between replacement tube and original tube at one end. After the joint is welded, snip off the pulling wire flush with the surface of the tube and weld over the drilled hole.

Allow this weld to cool, then adjust the brace to provide for contraction and shrinkage. After the brace is adjusted pull the sleeve into position and tack-weld the gap at the other end of the replacement tube. This will hold the joint in alinement. Remove the brace to eliminate any restraint of the contraction forces at this joint, then, finally, complete the weld around the gap previously tack-welded.

The replacement tube should now be welded into place— neatly reinforced with inner sleeves at the joints.

The same general method can be used for any of the other splice methods we will discuss.

A partial replacement tube of the same diameter as the original, reinforced by outside sleeves, is another method of

repairing tubing. Since this method requires the greatest amount of welding, there is more danger of distortion from the welding heat generated. Therefore, it is the least desirable of the methods.

There may be occasions, however, when the damage is so located that neither the inner sleeve type of splice nor the larger diameter replacement tube is applicable. In such cases, the outer sleeve type of splice, such as shown in figure 39, is the only alternative.

To accomplish the outside sleeve reinforcement, proceed I according to the following step-by-step outline.

First, cut out the damaged section of tubing by sawing straight across, not diagonally, making certain that the cuts are located away from the middle third of the affected area. Then cut a section of replacement tubing which matches the original tubing in diameter, wall thickness, and length.

A gap of not more than one thirty-second inch should exist between the ends of the replacement tubing and the stub ends of the original tubing.

For the outer reinforcement sleeves, select a length of tubing with an inside diameter equal to the outside diameter of the original tubing. A clearance of no more than one sixty-fourth inch should exist between sleeves and tubing.

Saw out the sleeves with either diagonal or fishmouth ends—fishmouth ends are preferable. Make the sleeves long enough so that their nearest ends are 1*4 tube diameters from the ends of the cuts in the original tubing. !

File off the burr from the edges of both sleeves and reinforcing tubing, then dip the sections in a bath of hot (165° F.) Paralketone. Wipe off the tube and sleeves, and slip the two sleeves over the replacement tubing. Aline the replacement tubing with the stub ends of the original tubing and push the sleeves along until they are centered over each joint. Rotate them to suit the space and to provide the greatest reinforcement.

Tack weld the two sleeves to the replacement tube in two places, then weld both ends of one sleeve to the replacement tube and original tube. Allow this weld to cool to prevent undue warping before welding around both ends of the other sleeve.

LARGER DIAMETER REPLACEMENT TUBE

Dispensing with reinforcing sleeves and using a larger diameter replacement tube requires the least amount of cutting and welding. It is thus desirable from the viewpoint of controlling distortion set up by the heat of the welding flame. This technique cannot be used, however, when a tube is damaged too near a cluster joint due to the fact that there will not be a sufficiently long stub end.

Carrying the splice past the joint into the next section of tubing is not recommended because it adds unnecessary weight to the structure. Nor can tubing of larger diameter be Used if brackets are mounted on the original tubing. In this latter case, the replacement tubing must be of the same diameter as the original.

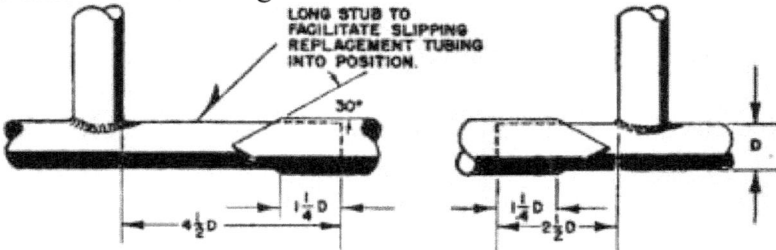

Figure 42.—Larger diameter replacement tube.

When this type of splice is used, cut out the damaged tubing so that a short stub of at least 2y 2 tube diameters is at one end and a long stub with a length of at least 4y 2 tube diameters at the other end. The cuts must be outside the middle third, as shown in figure 42.

Cut the replacement tube from a spare length of steel tubing having an inside diameter approximately equal to the outside diameter of the original tubing. The clearance between the two should not exceed one sixty-fourth inch. The fishmouth style of cut is preferable for the ends of the replacement tube.

The replacement tube should be sufficiently long that each end extends at least 1*4 tube diameters past the ends of the original tube.

File the burr from the edges of the replacement tube and the original tube stubs, then dip the replacement tube in hot Paralketone, wiping off the outside of the tube following immersion.

Spring the long stub of the original tube slightly out of normal position so that the replacement tube can be slipped over it. Then pull the replacement tube over the short stub and center it over the cut in the original tubing.

Tack weld one end of the replacement tube in four places, then quarter weld completely around that end. Allow the weld to cool, then weld the other end of the replacement tube to the original tube.

TEE AND CLUSTER JOINTS

Joints in which two tubes meet at an angle are quite common in aircraft construction. The simplest and most frequently used joint is the T-joint, illustrated in figure 43, which usually consists of an auxiliary member welded at right angles to a main or continuous member. If the two members join at an angle other than 90°, they constitute a saddle joint.

Proper construction of the T-joint involves only a small amount of fitting. This fitting can be done by filing the end of the vertical member to a concave shape and fitting it over the rounded surface of the horizontal member, as shown in figure 43.

Where the tube must be welded in a fixed position—as is likely to be the case when airplane structures are repaired— a combination of techniques will probably be employed. At

CUT TO A SNUG FIT WITH A RAT-TAIL FILE.

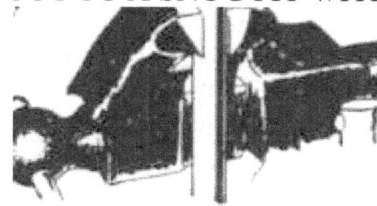

Figure 43.—T-joint.

some points on the tubing, the backhand method will give the desired control of the puddle, so that thorough fusion can be obtained. In a restricted area, it may be necessary to "dig in" with the torch to secure penetration to the root of the weld, even though this practice is frowned upon in the welding of flat stock.

When a number of tubes are welded at a common joint, the joint is referred to as a cluster joint. It is usually composed of a main member, a vertical tube, and other auxiliary tubes at varying angles to the main member. The vertical member is the first one attached. While it may be merely tacked in place, it is usually completely welded because of the additional strength. The other auxiliary members are then carefully fitted into place and welded.

In all cluster joints, the center lines of all members usually converge at a common point. By this precaution, stresses are distributed proportionately upon all members of the joint.

The term quarter welding, which has been mentioned several times in the preceding paragraphs, is a variation of skip welding used to avoid distortion. To perform this technique, first tack weld the tubing joint at four equidistant

places, then weld diametrically opposite segments from tack to tack.

AUXILIARY REINFORCEMENTS

Reinforcements such as gussets, inserts, wrapper gussets, and finger straps are often used in connection with tubular joints, especially the T and cluster joints, and in reinforcing clamps, such as those to which the engines are bolted. These reinforcements relieve some of the stresses on the joint itself, and increase the rigidity of the joint.

Gussets

Reinforcement plates in the form of flat gussets are often placed between the members of a cluster or T-joint to give added support. This principle is illustrated in figure 44 They are usually triangular in shape, with the legs of equal length.

The thickness of the gusset should be at least equal to the wall thickness of the tubing. It is usually welded on one side only. In many cases, the gusset does not extend into the apex of the angle but is instead notched out a short distance.

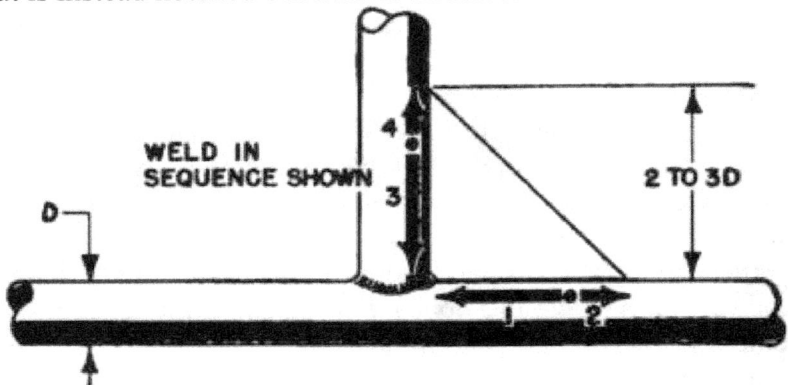

Figure 44.—Principle of and steps in w.lding tt» flat gusset.

There is considerable controversy as to the direction in which the weld should proceed, that is, whether it should begin in the restricted area or go toward it. The factor

determining the choice of direction for torch travel is the relative importance of limiting distortion caused by contraction when the weld cools, as opposed to the importance of preventing stress concentration in the joint. For example, if the job being welded is a portion of an engine mount which is not to be heat treated after welding to relieve stresses, welding from the restricted area out is probably more desirable.

If the unit is heat treated after welding to relieve stresses, the stress concentration is compensated for, and is not as important a factor as the prevention of distortion from contraction. It is then common practice to weld the first edge of the gusset in either direction, since the other edge is free to expand except for the tacks, which are placed near the ends of the seam. Welding of the second edge is completed by beginning at a point about one-fourth to one-half inch in from the outer edge of the gusset, and welding out to the edge; the remainder of the bead is then welded from the point near the outer edge to the inner part of the gusset, as explained in figure 44. Overlap the parts of the bead about one-eighth to one-fourth inch to secure thorough fusion. It is not advisable to lift the torch suddenly from the edge of the gusset, as a pinhole is likely to be formed which may serve as a point for the concentration of stresses, leading to failure of the member by cracking. Another recommended practice is to bring the bead around to the end of the gusset, rather than finishing it on the surface of the joint.

Inserts

A great deal of strength can be obtained from the reinforcement known as the insert. In this procedure, the tubes are slotted and filed to allow a snug fit over the plate which is inserted in the slot. The ends of the tube are filed to a convex shape, then hammered down with a special forming tool. This shaping is performed on both sides of the plate. The point at which an engine mount ring is welded together is

quite often a scarf butt with a plate inserted in the joint to give added strength at this point. Inserts are welded on both sides.

Wrapper Gussets and Finger Straps

The wrapper gusset is really a double gusset. It is made up of a square piece of material with side dimensions similar to those for ordinary gussets, as figure 45 describes.

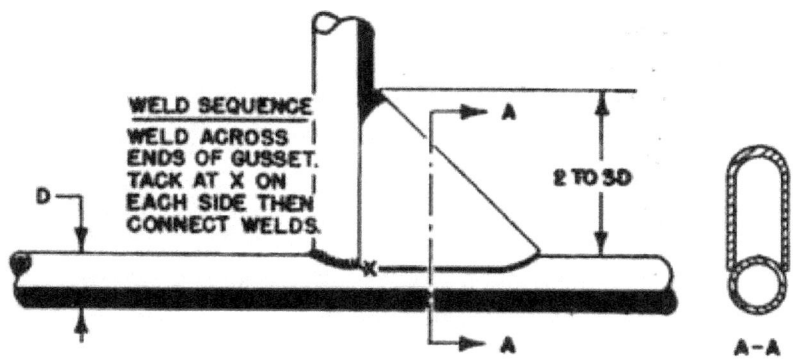

Figure 45.—Wrapper gusset.

The main member of any tubular assembly, with an injury such as a dent at points where truss members terminate, may be repaired by welding a sheet-metal patch over the damaged portion. This repair is known as the formation of finger straps. The metal is approximately the same thickness as the tube being repaired. The finger should have a width equal to the diameter of the brace tube, and a length equal to three or four diameters. The ends of finger straps and wrappers should be rounded to prevent the formation of an annealed zone in a straight cross-section of the tube.

SPACING

As noted in the sketches for the various types of tubing joints and splices, there may be two reasons for spacing the ends of tubing. Spacing may be necessary to allow for expansion, as in the case of the ends of the original tube inside

the outer sleeve splice. Spacing may also be required to allow penetration, as in the case of the ends of the original tube, where an inner splice is used. Fusion must extend into the inner sleeve, which cannot be accomplished without space between the ends of the outer members.

PENETRATION

Specifications for penetration will vary with the type of joint used. Butt joints, either plain, scarfed, or fishmouth, should have penetration to the full thickness of the tubing. Lap welds, such as those encountered in the various types of splices, have fusion including 100 percent of the thickness of the outer tube and from 25 to 50 percent of the inner tube. In the T-joint, penetration should extend from 25 to 50 percent of both members.

APPEARANCE OF BEAD

Beads may be either concave or convex in shape, depending upon the type of joint and upon the reinforcement required. Beads should be uniform in height and width, the ripples should be well defined and equidistant, and the edge of the bead should be well fused with the surface at all points.

MINOR REPAIRS

Most repairs on aircraft tubing require cutting-out and welding-in of a partial replacement tube, or replacement of an entire new section of tubing. Some repairs, however, are comparatively minor, an example of which is the straightening of a slightly buckled or bent piece of fuselage tubing.

Such repair is relatively simple, and if the part is made of chrome molybdenum in a non-heat-treated condition, it will actually be stronger after having been straightened. This strength is due to the cold-working the metal undergoes during straightening.

The equipment required consists of a steel screw C-clamp, three blocks of hardwood, and a piece of heavy iron beam the

same length as the bow, or bend, in the tube. Cut the wood blocks to fit the shape of the tube and line the grooves with leather or canvas. Place one of the grooved blocks at either end of the bent section and apply the beam so that it spans the bent area and backs up the two blocks.

Apply the third block on the opposite side of the tube at the point where the bend is greatest. Slip one end of the C-clamp over the iron beam and tighten the clamp down on the block at the center of the bend.

Tighten the clamp until the tube is bent slightly in the opposite direction, then remove the clamp and the block. Check the alinement of the tube by placing an accurate straightedge on both its side and top. If the straightedge check shows that a slight bow remains in the tube, reapply blocks and clamp and repeat the process until the tube lines up with the straightedge in both planes.

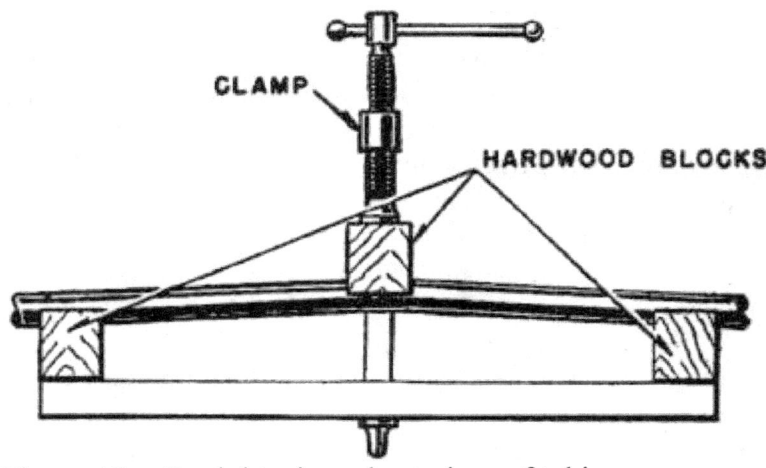

Figure 46.—Straightening a bent piece of tubing.

If cracks appear at the point where the maximum bend was corrected, drill a hole at the ends of the crack and weld a split sleeve over the crack. In each case where a bent tube is straightened, carefully test all nearby welded joints for cracks. When cracks appear, they must be repaired.

Let us suppose that a small, smooth dent is found in a length of tubing, but the piece is not out-of-round for any considerable length. A simple way of removing the dent is to push it out by air pressure. To accomplish this repair, remove one of the self-tapping screws provided at the ends of the main steel tubes, and apply an air pressure of 75 pounds or more per square inch to the inside of the steel tubing. With a torch, heat the dented area evenly to a dull red until the internal air pressure forces out the dent and restores the original shape of the tube.

If the combined internal air pressure and heat are not sufficient to remove the dent, tack-weld a welding rod to the center of the dent, and pull on the rod while heating the area. When the dent is removed, disconnect the rod, allow the area to cool, and then release the internal air pressure. Finally, replace the self-tapping screw which was removed.

When a crack is discovered in tubing, remove all finish from the area with steel wool or a wire brush. If the crack is located in an original weld bead, carefully chip, file, or grind on the existing weld bead, and reweld over the crack along the original weld line. Do not remove any portion of the existing tube or reinforcing material when grinding off the weld bead.

If the small crack is near a cluster joint but not actually on the weld bead, remove the finish and then drill a No. 40 (0.098) hole at each end to prevent the crack from growing. Then weld an overlapping piece of metal over the area. When the job is completed, apply a coat of zinc chromate primer to the area from which the finish was removed. Finally, apply finish coats to match the adjoining surfaces.

If previous attempts to straighten the tube have caused dents or cracks in its structure, it can be either reinforced or the damaged part can be cut out and replaced. If the damage is not too serious, reinforce the part by means of a split sleeve without removing the tube. In such event, the reinforcing

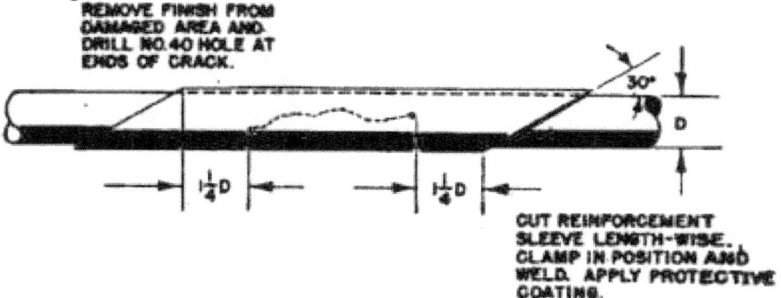

Figure 47.—Cracked-tube reinforcement by means of a split sleeve.

material should be a piece of tubing having an inside diameter equal to the outside diameter of the damaged tubing. Both tubings should have the same wall thicknesses.

Cut both ends of the reinforcing sleeve diagonally at about a 30° angle. The sleeve should be long enough to extend a distance of 1*4 tube-diameters past each end of the crack. Cut the reinforcing sleeve in half lengthwise, and separate the half sections. Remove the finish from the surface of the affected area with steel wool for 3 inches on each side of the damage. This done, clamp the two sleeve sections over the damaged area of the original tubing. Weld the two halves of the reinforcing sleeve together and then weld both ends of the sleeve to the damaged tubing. Refinish the surface of the joint according to instructions given earlier on anti-corrosion precautions.

QUIZ

1. What are the characteristics of chrome molybdenum (S. A. E. 4130) ?
2. What technique is used on chrome molybdenum?
3. What types of rods are used on S. A. E. 4130?
4. How do tubing joints differ from joints in flat stock?
5. What is a telescope joint?
6. How should tubes and sections be cut for repair work?
7. Why is a rosette sometimes added to a splice ?
8. What are inner reinforcing sleeves? How are they used ?
9. What is a T-joint ? A cluster joint?
10. What is a gusset? Wrapper gusset? Finger strap? Insert?

CHAPTER 7 CUTTING FERROUS METALS PRINCIPLES OF OXYACETYLENE CUTTING

There are numerous everyday uses for oxyacetylene flame cutting. It is a quick, inexpensive way to cut iron or steel where the effect of burning or heating the edge of a piece of metal is not objectionable. While the cutting of metals with an oxyacetylene torch is not done on aircraft, it has many uses in general service work around the shop.

Ferrous metals combine with oxygen so readily that the oxygen in the air can start the reaction, as rusty pieces of iron in scrap piles will attest. The rust is iron oxide, and the longer a piece is exposed to the elements, the more it is worn away and the more rust it collects.

Cutting iron or steel with an oxyacetylene torch is simply a speeding up of this process in a localized area, because iron oxidizes much more rapidly when it is hot. Pure oxygen, if

23*.-,nn° -.-,:! _io

directed on a hot piece of iron, increases the rate of oxidation so enormously that the metal is actually burned away.

The metal is heated to a bright red (1,600° F.), which is its kindling temperature, and a jet of high pressure oxygen is directed against it. This oxygen blast combines with the hot metal and forms an intensely hot oxide. The molten oxide is blown down the sides of the cut, heating the metal in its path to a kindling temperature. The metal thus heated also burns to an oxide which is blown away on the underside of the piece. Some unoxidized and partly oxidized metal is also removed by the force of the rapidly moving oxygen jet. This action is precisely that which the torch accomplishes when the mixing head is replaced with a cutting attachment or when a special cutting torch is used.

Figure 48 is an example of a cutting torch. It has the conventional oxygen and acetylene needle valves, which control the flow of the two gases. Many cutting torches have two oxygen needle valves so a finer adjustment of the neutral flame can be obtained.

A cutting torch combines a heating flame with a jet of pure oxygen under pressure. The heating flame preheats the metal to a bright red, and the oxygen jet is directed upon the hot metal to burn it away and thus form a slit known as a kerf, in the metal.

The heating flame in a cutting tip is generally not fed by a single hole as in a welding tip, but instead comes out through several holes which are arranged in a ring around a larger central hole for oxygen. The central oxygen tube tapers as it reaches the tip opening to increase the velocity.

The high pressure cutting oxygen jet is regulated by an auxiliary oxygen control valve generally operated like a trigger. This is shown in figure 48.

Four different sizes of tips are usually supplied for cutting metals of varying thicknesses. There are also special tips for cleaning metal, cutting rusty, scaly, or pointed surfaces, rivet washing, and so forth.

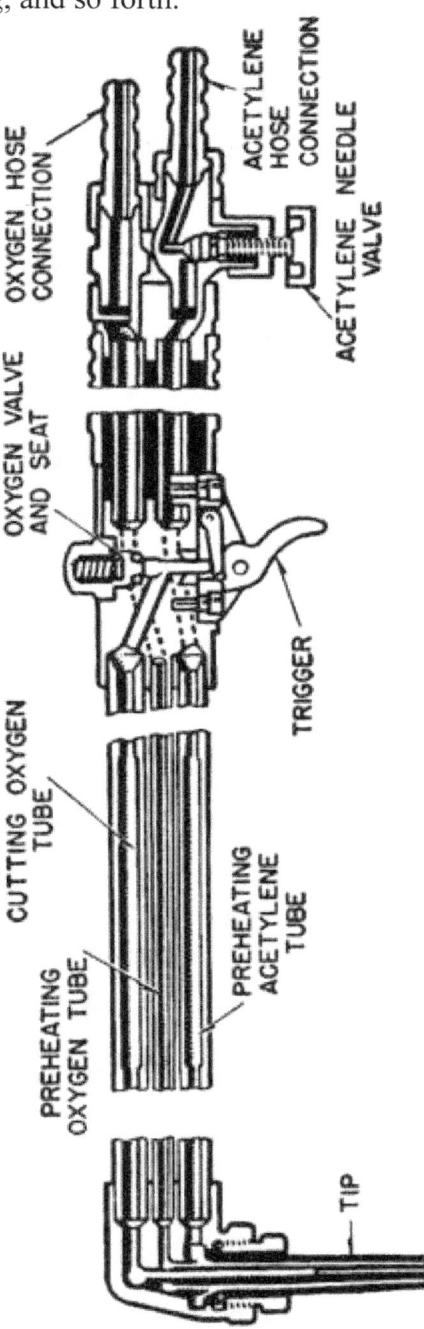

In cutting, as in welding, the pressure of oxygen and acetylene and the size of tip is determined by the thickness and quality of the metal to be cut. Table 11 shows the approximate pressure for various tip sizes. If the steel to be cut has a heavy coating of rust or scale, a greater oxygen pressure is needed to make the oxygen burn entirely through the metal.

Tobto 11 .—APPROXIMATE PRESSURE FOR VARIOUS TIP SIZES

Turn on the acetylene needle valve of the torch, light the gas, and adjust for a neutral flame, as in welding. The neutral flame is used in a cutting torch to bring the metal to a kindling temperature. In the case of plain carbon steel for example, this temperature is 1,400° to 1,600° F.

When the neutral flame is burning smoothly, pushing down on the triggerlike oxygen control lever will disclose the type

of cutting flame created. It may be necessary to readjust the neutral, preheating flame, when pressing the control lever, to make sure that it remains neutral. Let us attempt our lesson in cutting on a straight piece of metal. Draw a soap-stone line on the piece, about one-fourth to one-half inch from one edge. Then place the metal so that this line is beyond the edge of the welding bench. If an exceptionally straight cut is desired, clamp a bar of steel across the piece of metal to guide the torch.

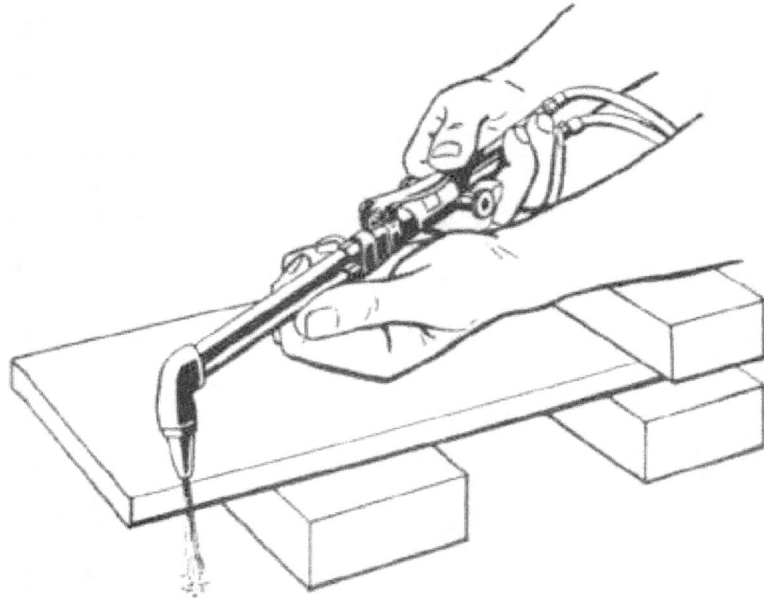

Figure 49.—Starling the cut.

With the preheating flame adjusted to neutral, grip the torch firmly, but not tightly, holding it so that there is instant access to the oxygen control lever. Figure 49 describes the proper method of beginning a cut in a piece of steel. Your left hand is used to guide the torch nozzle and the thumb of your right hand is used to press down the oxygen control lever.

Hold the torch steady. If the cutting tip wavers from side to side, a wide kerf which results in a rough cut, slower speed, and greater oxygen consumption will be made.

Begin cutting at the edge of the piece. Hold the tip vertical (at a right angle) to the surface of the metal, keeping the inner cone about one-sixteenth inch from the soapstone guide line. Hold the flame at this point until a spot in the metal turns bright red, then gradually press the oxygen control lever and move the torch steadily forward along the guide line to make a fast but continuous cut. A shower of sparks falling from the underside of the piece indicates that the cut is proceeding correctly.

If the cut does not seem to penetrate entirely through the metal, the cut is being made too rapidly. Release the oxygen control lever so that the pure oxygen is closed off, reheat the metal until it is bright red, and continue the cut.

If the cut is progressing slowly, the heat of the preheating flame remains too long in one spot, melting the edges of the cut, and resulting in a very ragged kerf.

When the cut is finished, the cut section may stick to the main piece. This means that some of the slag produced by the cutting action has bridged across the bottom of the two pieces and on cooling has formed a thin crust which holds them together. The crust is quite brittle, however, and a smart blow from a hammer will break it and separate the pieces.

SPECIAL CUTTING OPERATIONS

If the cutting job is on a piece where a cut cannot be started at the edge, but must begin within the piece, a longer period of preheating is required before cutting begins. When a spot is heated to bright red, raise the torch-cutting nozzle one-half inch from the work before pushing down the oxygen lever which turns the flame into a cutting flame. After a hole is cut through the metal, lower the torch to its normal position, one-sixteenth inch from the work, and then proceed.

When cutting a piece of round bar steel, such as shown in figure 50, begin at the side of the bar about 90° from the top, keeping the torch perpendicular to the piece. Gradually lift it to follow the circular outline of the bar as figure 50 demonstrates.

This vertical position of the torch is also necessary while descending the other side of the round piece.

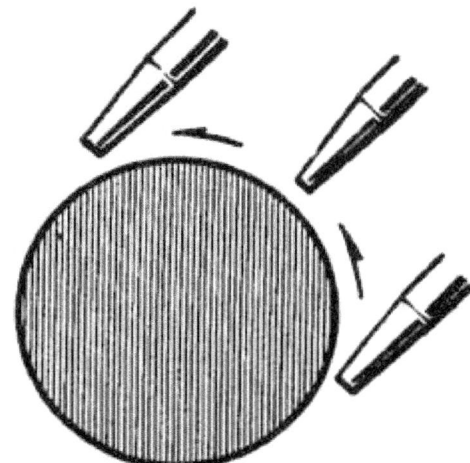

Figure 50.—Torch position for cutting round bar steel.

Figure 51 shows a piece of steel plate being beveled. Do not hold the torch vertically, but slant it at the correct angle for making the beveled edge. To obtain an even bevel, support the torch by resting the edge of the nozzle on the work, or guide it with a piece of angle iron clamped across the piece as figure 51 illustrates.

It is a relatively simple job to cut a hole in steel with a cutting torch. Preheat a small round spot in the steel to a bright red, then push down the oxygen lever and raise the torch slightly, A neat hole is thus pierced in the metal.

For larger holes, trace the shape with a piece of soapstone, then, at some point inside the circle, start a small hole. Work the cut out to the line and continue around the outline.

Figure 51.

veling steel with a cutting torch.

BEWARE OF SLAG

In all cutting operations, a firewatch should stand by with a CO_2 fire extinguisher. Globules of hot slag, if they lodge in combustible materials, may start a fire.

The drops of hot slag will roll along the deck for a considerable distance, and a space should be left clear for 30 to 40 feet around the job. If it is impossible to clear away materials that might catch fire, cover them with sheet metal guards or asbestos blankets. Be particularly careful about acetylene cylinders.

QUIZ

1. Describe an oxyacetylene cutting torch.
2. Give a step-by-step description of the procedure used in cutting metal.
3. Name several special cutting operations.
4. What precautions should be taken against fire when cutting metal ?

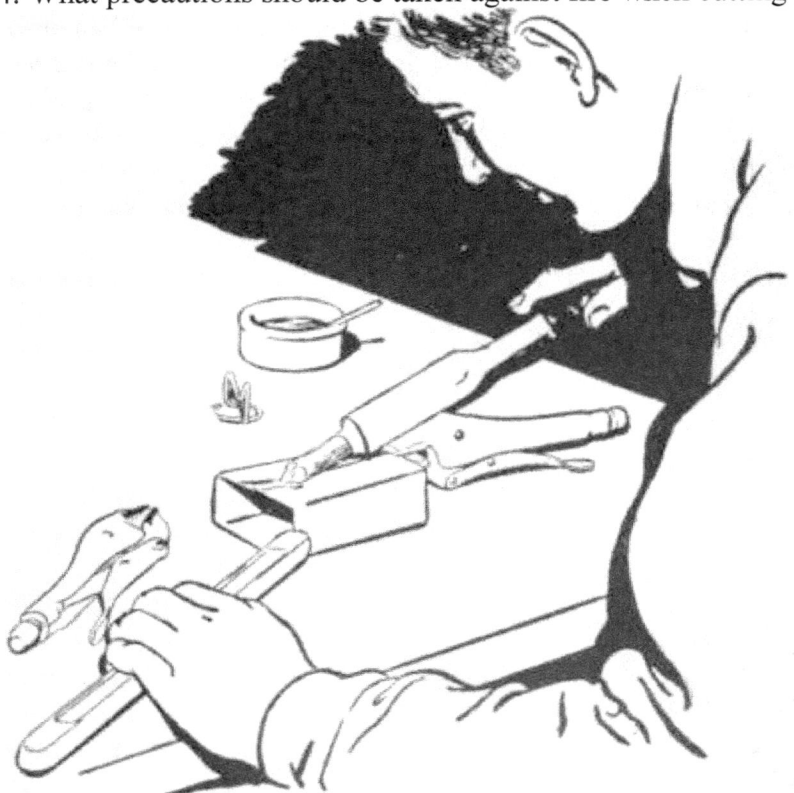

CHAPTER 8 BRAZING AND SOLDERING BRAZING

Brazing, originally, meant joining with brass, or "brassing." As the process was improved and new joining alloys were developed, the term assumed its present meaning—a group of thermal joining processes in which the bonding material is a nonferrous metal or alloy with a melting point higher than 800° F., but less than the metals being joined. Brazing, therefore, includes silver soldering, bronze welding, and hard soldering.

Brazing requires less heat than welding, and therefore may be used to join metals that are injured by high heat. The strength of brazed joints is not so great as welded joints, however, and, for this reason, brazing is not used for structural repairs on aircraft.

As the definition of brazing implies, the base metal parts are not melted. The brazing metal adheres to the base metal by molecular attraction; it does not fuse and amalgamate with them.

The usefulness of brazing is easily recognized when the many metals that can be joined by this process are considered. Brazing is applicable to the joining of cast iron, malleable iron, carbon steels, alloy steels, wrought iron, galvanized iron and steel, copper, and brass and bronze and nickel alloys. It is also used to join dissimilar metals, such as cast iron to steel, or steel to copper.

Its principal use is in maintenance, making and repairing tools, jigs, and machinery. In this field it has many applications and attendant advantages. Among these are the relatively low temperatures involved, reduced chance of an excessive annealed area near the brazed joint, and the ability of a properly prepared brazed joint to stand heavy compression and impact loads.

BRAZING TECHNIQUE

In brazing a joint, first bevel down the edges as in welding steel. Clean the surrounding surfaces of dirt, rust, and so forth, then select the proper brazing alloy for the job.

The Navy supplies four brazing alloys which are graded A, B, C, and D. Grades A and B with melting points of about 1,600° F., are used for strong connections on steel, cast iron, brass, bronze, and general brazing of nickel alloys. Grades C and D are suitable for brazing steel parts that are to be subjected to heat treatment under 1,600° F., after brazing. The melting point of grade C varies from 1,650° to 1,760° F., and that of grade D from 1,725° to 1,825° F.

Because of its higher melting point, grade D brazing alloy
is more difficult to apply than grade C, but is preferable with chrome-vanadium and chrome-molybdenum steel. When selecting a brazing alloy, choose one whose melting point is at least 100° F. less than the metal being joined.

A brazing flux is necessary to obtain a good union between the base metal and the filler metal. A good flux for brazing steel is a mixture containing two parts borax and one part boric acid. Use a neutral torch flame and move it with a slight, semicircular motion.

Preheat the base metal slowly with a mild flame, and when it reaches a cherry red heat (in the case of steel), heat the rod and dip it into the flux. Enough flux adheres to the rod and it is not necessary for you to spread it over the surface of the metal.

Bring the filler rod near the tip of the torch and let the molten bronze flow over a small area of the seam. Remember, the base metal must be at the flowing temperature of the filler metal before it will flow into the joint. The brazing metal melts when applied to the steel and runs into the joint by capillary attraction. Continue adding the rod, as the braze progresses, with a rhythmic dipping action so that the bead will be built to a uniform width and height. Complete the job rapidly and with as few passes of the rod and torch as possible.

The ideal brazing job is completed in one pass. Avoid multiple layers, and if the job requires more than one pass, always remove and replace the spent flux before applying succeeding layers of filler metal.

It is important that the brazing temperature be carefully controlled. If the base metal is heated excessively above the flow temperature of the brazing alloy, the bronze will boil when added and the low melting point alloys of the bronze will burn out, leaving the bronze porous and brittle. On the other hand, if the base metal is not hot enough, the bronze will not flow smoothly, but will form elusive drops which roll off as fast as the bronze is applied.

After finishing the job, allow it to cool slowly.

BRAZING CAST IRON TO STEEL

To braze cast iron to steel, flow molten bronze from a filler rod over the hot surface of the metal to be joined to obtain a solid bond between the edges of the seam.

Heat the work to be joined to a temperature slightly above the flow point of the brazing alloy. In the case of steel, this point is determined when the metal reaches a cherry red color. Metals which lose their original qualities when melted can thus be joined by brazing without undergoing loss of those qualities.

Brazing is usually the best method of joining unlike metals (copper and steel) or two like pieces of a metal such as malleable cast iron which has been heat-treated.

Since the base metal is not melted in brazing, the joining process is greatly simplified. The preheating necessary in fusion welding is largely eliminated.

In deciding whether brazing of a joint is justified, remember that a metal which will be subjected to a sustained high temperature in use should not be brazed.

BRAZING BRASS

Brazing, rather than welding, is the most effective method of joining brass, because such technique requires a filler rod with a melting point slightly lower than the base metal. Thus melting of the base metal is eliminated.

Brass, in its simplest form, is an alloy of copper and zinc, although other metallic elements are often added to improve its characteristics.

Naval brass, one of the best of these alloys, consists of 62 percent copper, 0.5 to 1.5 percent tin, 0 to 0.10 percent iron, 0.20 percent lead, and the remainder zinc. Its three advantages are high strength, toughness, and resistance to corrosion. It comes in bar, plate, rod, sheet, and strip form, and in soft, half-hard, and hard condition.

Brass has few uses in aircraft, other than in pipe fittings, and, consequently most of the brazing will be in connection with general repair Work on shop equipment, not on aircraft.

The torch flame with brass should have a slight excess of oxygen—one of the very few instances where an oxidizing torch flame is used. Be especially careful in applying heat to brass to avoid burning or oxidizing the zinc content of the brass.

Any good commercial brazing and welding flux will do for this operation, or, in an emergency, borax diluted with boric acid or sodium carbonate may be used. Apply the flux by dipping the hot end of the filler rod in the mixture or by painting the dissolved flux on the rod. Flux protects the hot metal from the air and other gases by forcing a film over it, and also cleans the hot brass of oxides formed during the welding process.

The filler rod for brass should have approximately the same composition as the base metal. As pointed out earlier, a rod with a slightly lower melting point than the base metal gives the best results. Use either a grade A or B rod for this job. Two commercial filler rods which are also good are Tobin bronze and manganese bronze rods.

The joints used in brazing brass are the same as those used in welding.

Be sure to clean the surface with a file or abrasive cloth, and allow for expansion and contraction. Thick pieces of brass must be beveled by filing or by some other mechanical method. Never bevel brass by melting or cutting since this method destroys the zinc in the brass. To reduce the amount of heat required for the actual brazing and lessen the danger of warping, preheat heavy brass parts.

SILVER SOLDERING

Silver soldering is one of the several methods of brazing. Silver base solders are used in aircraft work to repair oil coolers, coolant radiators, and other parts which must withstand vibration and high temperatures. Silver solder is used extensively to join copper and its alloys, monel metal nickel, and silver, as well as various combinations of these metals, and thin steel parts.

Table 12.—SILVER SOLDERS

Class	Silver[1] (percent)	Copper[1] (percent)	Zinc[1] (percent)	Other[1] (percent)	Melting point (°F.)	Flow point (°F.)	Uses
0	20	45	35	....	1,430	1,500	Ordinary brazing purposes where higher physical properties are required than provided by brazing alloys and where service or appearance does not require a high silver content.
1	45	30	25	....	1,250	1,370	Intended for the general range of silver soldering.
2	65	20	15	....	1,280	1,325	High silver content used where high strength, resistance to corrosion, and good appearance is required.
3	15	80		[3] 5	1,200	1,300	Intended for brazing copper and copper base alloys. Do not use for ferrous alloys.
4	50	15	16	[2] 19	1,160	1,175	General purpose intended for brass, copper, ferrous metals and particularly nickel-copper alloys and alloy steels.
5	50	15	15	[2] 17 [4] 3	1,195	1,270	Same as class 4, but where design requires addition of a fillet or where close tolerance cannot be maintained and filling is necessary. Also for hard materials such as cemented carbides for tools.
6	50	15	25	[2] 10	1,166	1,190	Same as class 4.

[1] Approximate percent. [2] Cadmium. [3] Phosphorus. [4] Nickel.

Silver solder can be obtained in several different grades with silver content ranging from 14.5 to 66 percent, and melting points varying from 1,160° to 1,430° F. The standard forms of silver solder are strips and wires, but it is also made in rod form.

If the job to be performed is one in which the solder may be placed in the joint before applying heat, use the strip form. For joints requiring the solder to be applied after heating, use the wire form.

It is necessary to use flux in all silver soldering operations because of the necessity for having the base metal chemically clean without the slightest film of oxide to prevent the silver solder from coming into intimate contact with the base metal

A paste flux is used generally in most silver soldering. If a prepared flux is not available, a mixture of 12 parts of borax and 1 part boric acid is satisfactory for high-melting-point silver solder. Prepared flux begins to melt at 800° F., becomes fluid at 1,100° and remains stable up to 1,600° F. It melts at a slightly lower temperature than the solder.

The joint must be physically clean, which means free of all dirt, grease, oil and paint, and also chemically clean— minus all traces of oxide film. After removing the dirt, grease, and paint, remove any oxide which may be present by grinding or filing the piece until bright metal may be seen. During the soldering operation, the flux continues the process of keeping oxide away from the metal.

Joints to be silver soldered must have smooth edges and must fit tightly together. Only a film of silver solder is usually needed for a sound joint. Strength is not added to

LAP JOINT FLANGED BUTT JOINT EDGE JOINT

rrmMm^ rmTTi^rmTm vzzzzzzzM

"^SOLDER ^SOLDER SOLDER

Figure 52.—Recommended joints for sllvor soldering.

the joint and expensive solder is wasted if it is used as a filler metal.

In figure 52 are presented three recommended types of joints for silver soldering. Flanged, lap, and edge joints, in which the metal may be formed to furnish a seam wider than the base metal thickness, furnish the type of joint which will bear up under all kinds of loads. If a lap joint is used, figure the amount of lap according to the strength needed in the joint. Here is a handy rule of thumb: For strength equal to that of the base metal in the heated zone, the amount of lap should be four to six times the metal thickness for sheet and small diameter tubing.

The oxyacetylene flame for silver soldering should be neutral, but may have a slight excess of acetylene. It must be soft, not harsh. During both preheating and application of the solder, hold the tip of the inner cone of the flame about one-half inch from the work. Keep the flame moving so that the metal will not be overheated.

If the piece is large preheat a considerable area around the joint before applying the solder, especially if the base metal conducts heat rapidly. When soldering two pieces which have different thicknesses, or which conduct heat with unequal speed, gage the preheating so that both parts reach the soldering temperature at the same time.

When both parts of the base metal are at the right temperature (indicated by the flow of flux), begin applying solder to the surface of the under or inner part at the edge of the seam. It is necessary to simultaneously direct the flame over the whole seam and keep moving it so that the base metal remains at an even temperature.

SOFT-SOLDERING

Soft-soldering is used chiefly for copper, brass, and coated iron in combination with mechanical seams—that is, seams that are riveted, bolted, or folded. It is also used where a leakproof joint is desired, and sometimes for fitting joints to promote rigidity and prevent corrosion. Soft-soldering

is generally performed only in very minor repair jobs. This process is also used to seal electrical connections. It forms a strong union with low electrical resistance.

Soft solder yields gradually under a steadily applied load and should not be used unless the loads transmitted are very low. It should never be used as the sole means of attachment of two structural members.

The Soldering Copper

A soldering copper, sometimes incorrectly referred to as a soldering iron, is the tool used in soldering. Its purpose is to act as a source of heat for the soldering operation. The bit or working face is made from copper, since this metal will readily take on heat and transmit it to the work.

The bit should be relatively blunt. If it is too thin and ' pointed it will cool too rapidly. Figure 53 shows a correctly shaped bit.

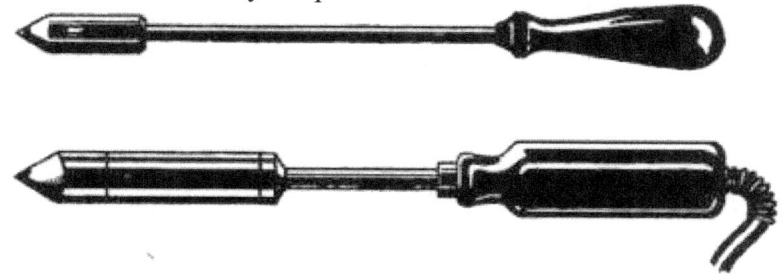

Figure 53.—Soldering coppers—common and electric.

Soldering coppers may either be heated by blowtorch, gas flame, or electricity. Electric soldering coppers have an internal heating element and are especially useful where a small but constant heat is required, as in soldering electrical connections.

Tinning the Copper

To tin the copper, first heat it to a bright red and then clean the point by filing until it is smooth and coppery.

238800 c 53 11

No dirt or pits sfhould remain on its surface. Clean the point chemically by dipping it into a cleaning compound while it is still hot This removes all the oxides. Then apply the solder.

These two operations may be combined by melting a few drops of solder on a block of sal ammoniac (cleaning compound) and then rubbing the soldering copper over the block until the tip is well coated with solder. A properly tinned copper has a thin unbroken film of solder over the entire surface of its point.

If the point of the copper needs reshaping, it should be done by forging. In performing this step, remove all of the oxides and reheat the copper to a bright red. Using a heavy hammer, forge the point to the desired shape on an anvil, then tin in the usual manner.

When using the copper, occasionally dip the point in a solution of one part of sal ammoniac to 30 parts of water. (Keep this solution in an earthenware jar.) If sal ammoniac is not available, powdered rosin will serve to tin the point.

The fluxes ordinarily used for soft soldering are solutions or pastes containing zinc chloride (cut acid). The liquid or paste medium holding the flux material is evaporated by the heat of the soldering operation, leaving a layer of flux on the work. At the soldering temperature, the flux is melted and partially decomposed with the liberation of hydrochloric acid. This acid then dissolves the oxides from the solder and the work. The melted flux also forms a protective film on the work to prevent further oxidation. Zinc chloride flux is used on iron, copper, brass and galvanized iron.

Because zinc chloride fluxes have a corrosive action, it is necessary to employ a noncorrosive flux for work on electrical connections or other places where all traces of flux cannot be removed. Rosin, either in powder or paste form, is the most commonly used flux of this type.

Use a noncorrosive commercial flux or a rosin flux for electrical connections or where it is impossible to remove all traces of flux from the work. Zinc chloride and other corrosive fluxes must be washed from the work to prevent corrosion.

Types of Soft Solder

Soft solders are chiefly alloys of tin and lead. The percentages of tin and lead vary considerably in various solders, with a corresponding change in their melting points, ranging from 293° to 592° F. "Half-and-half" (50-50) solder is a general purpose solder and is most frequently used. It contains equal proportions of tin and lead and melts at approximately 360° F.

Soft solders are usually supplied in bar or wire form, although they can also be obtained in pig or granulated form for jobs requiring large quantities of solder. Some wire-shaped solders have a core of flux. The most popular solder of this type has a core of rosin and is used chiefly on electrical connections.

The film of solder between the surfaces of a joint must be kept thin to make the strongest joint.

Soldering Technique

The application of the melted solder requires somewhat more care than is apparent. The parts should be locked together or held mechanically or manually while tacking, as shown in figure 54. To tack the seam, touch the hot copper to a bar of solder, then use the drops of solder adhering to the copper to tack the seam at a number of points.

Hold a hot, well-tinned soldering copper so that its point lies flat on the metal at the seam, while the back of the copper extends over the seam proper at a 45° angle, and touch a bar of solder to the point. As the solder melts, draw the copper slowly along the seam as in figure 55. Add as much solder as necessary without raising the soldering copper from the job. The melted solder should run between the surfaces of the two sheets and cover the full width of the seam. Make long strokes, and when the copper cools, reheat the tool. Resume your work by remelting the solder where the operation stopped, and go on from there. The best seam, however, is produced without removing the soldering copper from the surface of the work.

In another type of soldering known as sweating, both surfaces of the pieces to be joined are tinned, then held together and heated with a soldering copper or blowtorch until the solder melts and begins to run out. Keep the parts in close contact by pressure until the solder cools and sets.

Figure 54.—Holding the parts together.

Post-Soldering Treatment

Whether doing hard soldering or soft soldering, the joint must be cleaned when finished. This means the removal of all flux that might cause corrosion or prevent paint from adhering. In some cases, immerse the joint in a "bright dip" to restore the color.

>

If the base metal is nonferrous—that is, not made from iron ore—a good solution for removing flux consists of 1 fluid ounce sulfuric acid, 1.5 ounce sodium bichromate, and 1 gallon of water.

Boil ferrous metals in a 10 to 15 percent solution of caustic soda for 30 minutes to eliminate the flux. In either case, rinse the metal thoroughly in clean water after treating it.

The job may require that metal which has been discolored by heat, in the soldering process, be restored to its original color. The color of copper and brass particularly seems sensitive to heat. The remedy is a "bright dip" consisting of 68 fluid ounces sulfuric acid, 20 fluid ounces nitric acid, 0.12 fluid ounce of hydrochloric acid, and 40 fluid ounces of water. Following this bath, rinse the metal thoroughly in clean, running water.

Scale which is caused by the heating of steel parts may be removed by a light sand blast.

Figure 55.—Soldering a seam.

QUIZ

1. Define brazing. !
2. What are some of the metals which can be joined by ! brazing?
3. Describe the steps in brazing cast iron to steel.
4. Describe the steps in brazing brass.
5. For what is silver soldering used ?
6. Describe the condition of a joint which is correctly pre- pared for silver soldering.
7. For what is soft soldering most often used by the Avia- tion Structural Mechanic ?
8. Should soft solder be used as the sole attachment of two structural members?
9. Describe the process of tinning the copper. j
10. Of what is soft solder composed ?
11. What must always be done with work being soldered? !
12. What treatment should the joint be given after solder- ing?

CHAPTER 9

ELECTRIC ARC WELDING
PROCESSES AND ELECTRODES

Since electric arc welding is a subject of broad scope, only the processes, equipment, and procedures common to the Aviation Structural Mechanic are discussed in this book.

The development of fusion welding by the electric arc welding process has been due to the practical application of the following phenomenon: When an arc is formed across a gap in an electrical circuit, an intense heat is instantly generated. The arc welding process uses this arc to melt the metal being welded and the electrode to add metal in the case of the metallic arc welding process.

Three types of electric arc welding—carbon arc, bare metallic arc, and the shielded metallic arc—make up the most common of the arc welding processes.

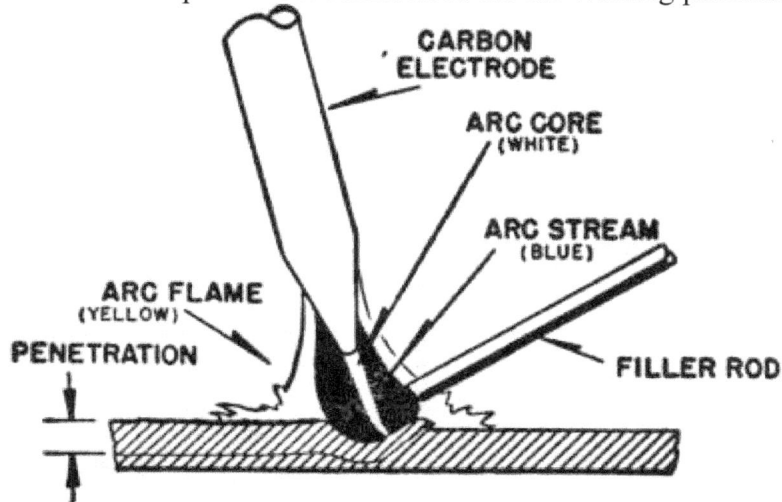

Figure 56.—Carbon arc.

CARBON ARC

The carbon arc is formed between a carbon electrode and the base metal—the metal being welded. In figure 56 you may see the component parts of the carbon arc.

The arc develops an intense heat which has a temperature range of from 7,000° to 9,000° F. This heat is used to melt the base metal. If additional metal must be added to complete the weld, a filler rod must be used just as in oxyacetylene welding.

Carbon arc welding is similar in many respects to oxyacetylene welding. However, the more intense heat of the carbon arc makes it necessary to manipulate or move the arc much more rapidly and without the puddling effect obtained in oxyacetylene welding.

The carbon arc may be used for welding when no metal is added; for welding with a filler metal; for preheating; and, to a limited extent, for cutting purposes.

METALLIC ARC

Although there are many applications where carbon arc welding may be used to advantage, fully 95 percent of all arc welding is done with the metallic arc. Because of this predominant use of metallic arc welding, we will devote considerable time to metallic arc welding processes.

As already stated, metallic arc welding is divided into two types—bare metallic arc, and shielded metallic arc.

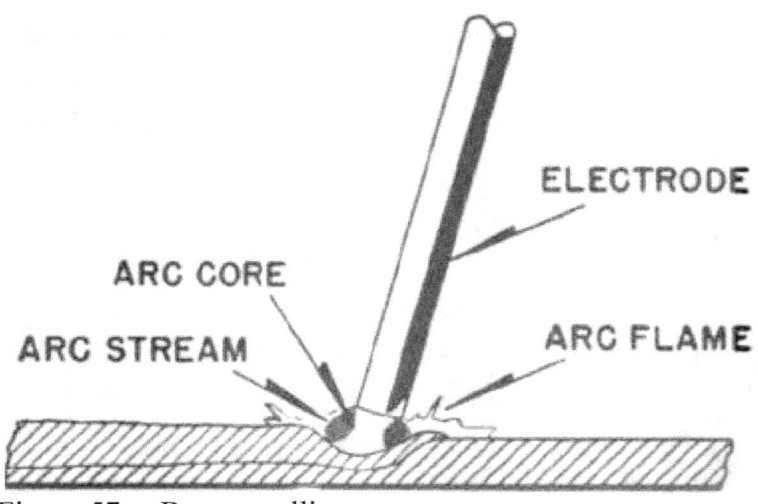

Figure 57.—Bare metallic arc.

Bare Metallic Arc

The bare metallic arc welding process uses the heat of the arc to melt the base metal and the wire or rod used sis an electrode. The molten metal of the electrode is carried across the arc and deposited as additional filler metal to complete the weld. A comparison of figures 56 and 57 will readily disclose that the component parts of the carbon arc and the bare metallic arc are the same; filler metal must, of course, be used with the carbon arc. The temperature of this arc is estimated to range from 4,.")00° to more than 7,000° F.

The bare metallic arc—this includes dusted, wiped, or lightly coated electrodes—affords little if any protection from the atmosphere to the molten metal passing from the electrode, and to the molten base metal. Consequently, the molten metal absorbs nitrides and oxides from the atmosphere and loses important elements by vaporization. This results in a weld that is poor in appearance and low in ductility and resistance to fatigue and impact. The weld, however, has a most satisfactory tensile strength.

Although the bare metallic arc welding process is used in comparatively few applications in aircraft maintenance and repair work, its fundamental principles are included here so that the welder may learn its use and manipulation for two reasons:

1. It is more difficult to hold an arc and to become a good welder with this sensitive type of electrode than it is with the shielded metallic arc type of electrode. Consequently, a welder who becomes proficient in welding with bare electrodes and masters their sensitive operating characteristics can readily learn to use the shielded arc electrode and will be the better welder because of such ability.

2. With the bare or dusted electrode, the beginning welder can see what actually takes place in the arc. He can thereby become familiar with the action occurring at the arc during the welding process.

Shielded Metallic Arc

The use of a coated or covered metal electrode in the arc welding process eliminates all the disadvantages of welding with a bare electrode. It is vitally necessary that we understand the difference between a so-called bare electrode and a covered electrode. For all practical purposes, an electrode which has a light coating applied by dusting, dipping, or washing, in order to stabilize the arc, is known as a bare

electrode. A coated or covered electrode is one that is covered with a thick material that serves the double purpose of stabilizing the arc and improving the properties of the weld metal.

The shielded metallic arc type of electrode, shown in figure 58, is by far the most widely used type of electrode. As can be seen in figure 58, this type of electrode shields the molten metal which passes through the arc and the molten base metal.

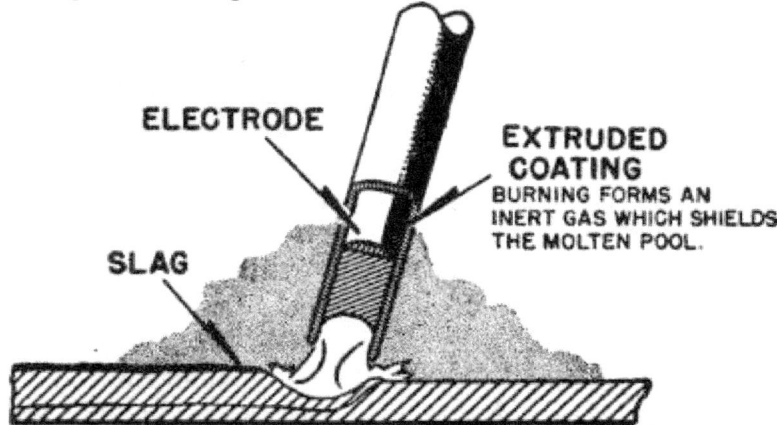

ELECTRODE

EXTRUDED
COATING
BURNING FORMS AN
INERT GAS WHICH SHIELDS
THE MOLTEN POOL.

SLAG

Figure 58.—Shielded metallic are.

In addition, the constituents of the electrode coating merge with the weld metal and form a slag coating over the weld. The maintenance or addition of the elements by the electrode coating results in a weld which is much more ductile, has greater tensile strength, and is comparable to the base metal in its physical properties.

The slag covering has several important functions during the welding process:

1. Removes oxides and impurities.
2. Retards rate of freezing (solidification) of metal.
3. Slows cooling rate of solidified metal.
4. Controls the shape and appearance of deposit.

ARC FORCE

Before using the arc it is necessary to know exactly how it functions. Along with melting the electrode and base metal and their subsequent fusing, the arc has the additional and important ability to dig. This digging quality of the arc is called arc force.

There is a definite force in the arc stream just as there is in a stream of water from a hose. As water forced through a nozzle will dig away dirt, so can the force of the arc stream be used to dig into the parent metal.

The important factor to observe in the utilization of arc force is causing the arc to travel at a sufficient speed to take advantage of the penetrating power of the arc force. This can be further illustrated by the comparison of water flowing from a hose.

The action of digging away dirt with a stream of water is only effective when the stream is directed at the dirt itself—not when directed into the pool of water which soon accumulates. If the stream of water is to keep .digging, it must keep moving fast enough to stay ahead of the pool.

The same reasoning can be applied to welding. When the arc is advanced too slowly, a pool of molten metal forms beneath it, and the force of the arc is expended in the molten pool instead of penetrating into the base metal at the root of the joint. This molten metal flows along the joint under the arc and tends to solidify in the root of the weld without fusing to the base metal,

When the arc is advanced at the proper rate of speed, the arc force digs into the base metal and the result is good penetration. However, at a slow rate of arc speed travel, there is usually a small puddle of molten metal under the arc which dissipates the arc force and prevents maximum penetration. The limiting speed is usually the highest speed at which the surface appearance remains satisfactory.

An increase in current increases the arc force and penetration just as the increase in the volume of water through

the same size nozzle increases the digging power of the stream of water. To use higher currents, larger size electrodes may be needed. In general, the first indication of excess current will be poor surface appearance of the weld.

In a further comparison of the arc to a stream of water from a hose, it is obvious that to dig deep into the dirt the nozzle must be kept near the ground to avoid spreading the stream of water into an ineffective spray.

When a long arc is held, heat is dissipated into the air, the stream of molten metal from the electrode to the work is scattered in the form of splatter, and the arc force is spread over a large area resulting in a wide, shallow bead instead of a narrow one with deep penetration,.

The degree of penetration is proportional to the current used, combined with the effectiveness of the use of arc force. An increase in current will increase the arc force and penetration. However, if the speed of travel is too slow, the arc force will be wasted and there will still be less penetration than could be obtained by taking full advantage of the force. In order to effectively use arc force for penetration, the speed of travel should be fast enough so that the electrode tip will be just ahead of the molten pool at all times, giving the arc force full opportunity to dig deep into the root of the joint.

POLARITY

To understand the meaning of polarity, its relationship to the arc welder, and its use, it will first be necessary that we become familiar with the basic principles of an electric current.

Electric current is divided into two classifications—direct current (d. c) which is electricity flowing in one direction only, and alternating current (a. c), or current flowing alternately in opposite directions. In the latter, the direction of flow is continually reversing itself.

Electric current consists of a movement of electrons through an electrical circuit; the movement being from the

negative terminal of the battery or generator through the electrical circuit to the positive terminal or pole of the source of electromotive force.

In a direct current circuit the polarity always remains the same and the current always flows through the circuit in the same direction; however, in an alternating current circuit the current continually reverses direction in keeping with the changes in polarity at the terminals of the a. c generator.

The polarity of the voltage drop across any electrical circuit, whether it be an arc, a single resistor or an entire circuit, will be determined by the direction of electron flow (current). The sum of all the voltage drops around the circuit equals the potential developed at the terminals of the generator.

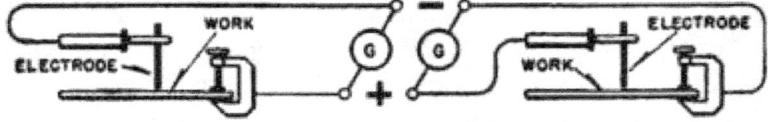

STRAIGHT REVI Flgur# 59.—Straight and rtvwit polarity.

When these principles are applied to electric arc welding, it becomes important for the arc welder to understand them thoroughly. This can best be done by considering the very first phases of arc welding.

There was a time when nearly all welding was performed with direct current and bare metallic electrodes. Under these conditions it was found desirable to connect the positive side of the arc to the work and the negative side to the electrode. This applied a greater portion of the heat to the work, since the positive side of the arc releases 65 to 75 percent of the heat. With the electrode connected to the negative side of the arc, the polarity of the circuit was said to be straight, as illustrated in figure 59.

Different jobs required different welding techniques, however—such as when welding cast iron or nonferrous metals—

and it became important to decrease the amount of heat in the work metal. This was achieved by connecting the work to the negative side and the electrode to the positive. Since the connection had been reversed, the direction of current was reversed, and the welding current polarity was said to be reversed, as may be seen in figure 59.

In the early days of arc welding, it was necessary to change the cable connections at the machine or at the work and electrode, if a change in the polarity was desired. With the advent of shielded metallic arc welding, the change in the polarity to meet the conditions set up by the different coated electrodes became more frequent. Coated electrodes are manufactured for specific jobs, and consequently some of them give better results with straight polarity than reversed polarity, while others produce satisfactory results when used with either polarity.

Manufacturers of welding machines soon realized that changing the cable connections to change the polarity for each job with different requirements was a cumbersome method. They therefore designed a polarity switch and mounted it on the control panel, thus making changing polarity a simple, easy, and quick operation.

The polarity to use with a particular electrode is established by the manufacturer. If doubt exists as to the polarity of the welding current at the electrode holder and ground clamp, simple checks can be made.

One way to check polarity is with a voltmeter. When the voltmeter connection marked positive (+) is connected to the ground lead and the other connection attached to the electrode holder lead, the needle will register the voltage if the leads are connected for straight polarity. If the voltage indicator needle drops below zero, the leads are connected for reverse polarity. Remember that in straight polarity, the electrode is negative.

If a voltmeter is not available, the polarity may be checked with a carbon electrode. If the electrode is negative (straight polarity), an arc can be established between the

end of a carbon electrode held in the electrode holder and a steel plate connected to the positive lead. This lead may be moved rapidly over the plate or drawn out to considerable length without being extinguished. However, if the electrode is positive (reversed polarity), the arc will be very unstable, and cannot be drawn out without becoming extinguished. If sharpened to a point, the electrode will become blunt and a soot deposit will form on the plate.

Perhaps the quickest way to determine polarity is to use an E6010 electrode. This electrode is very sensitive to polarity and is designed to be used with reversed polarity. It gives such greater results when used correctly that polarity of the circuit can be determined by merely observing its performance.

The information on polarity applies only to direct current With alternating current, since polarity changes very rapidly (120 times a second), the electrode must be suitable for use with either polarity. This does not mean, however, that a negative or positive polarity electrode will not be satisfactory with alternating current. As a matter of fact, a shielded arc electrode of high quality will give satisfactory results on alternating current, even though it is designed for use with direct current.

ARC BLOW

Since a magnetic field is set up around any conductor through which electric current is flowing, arc blow is a result of the combined influences of these fields around the arc, the electrode, and the work metal. Arc blow is most likely to occur when welding heavy metal in corners because of the proximity of the sides of the electrode to the metal being welded. Also, arc blow may cause uneven burning of the electrode coating, which in turn will result in improper fusion.

There are a number of methods by which arc blow may be reduced, minimized, or eliminated. Among these are the following:

L Reverse the direction of welding.

2. Weld toward a heavy tack or toward a completed weld.

3. Use back stepping on long welds.

4. Change the position of the ground.

5. Wrap the ground cable around the work several times.

6. Use double grounds, one at each end of the weld.

7. Place ground connection as far from the joint as possible.

If arc blow is encountered try these various methods or a combination of methods until a well-behaved arc is achieved.

Arc blow is much less noticeable when welding with alternating current than with direct current. As has been stated previously, this is due to the continuous change in the direction of current flow resulting in very little magnetic action with alternating current.

ELECTRODE CLASSIFICATION

Electrodes used to weld carbon and alloy steels are classified in the publication, Tentative Specifications for Iron and Steel Arc Welding Electrodes, published by the American Welding Society (A. W. S.), and American Society for Testing Materials (A. S. T. M.). In table 13, the outstanding features of this specification regarding the electrode classification number, the welding positions in which the electrodes may be applied, a general description mentioning the type of covering, current, and polarity recommended, are given for each.

A study of this chart shows that there are five major arrangements of electrodes in the heavily covered category, ranging in tensile strengths from 60,000 to 100,000 p. s. i.

To understand the significance of classification numbers, consider the E6010 group. The E represents tihe word electrode. The first two numbers—60—refer to the minimum tensile strength in the stress relieved condition, or 60,000 p. s. i. The third number explains the possible welding

positions, such as "1" for all welding positions (flat, vertical, overhead, and horizontal); or "2," which designates a greater restriction in choice by being usable only in the horizontal fillet and flat positions; whereas a "3" in the third number indicates that these electrodes may be applied in the flat position only. The fourth number in the classification indicates subgroups, and may be either "0," "1," "2," or "3" under the present system.

A. description of the welding characteristics of the electrodes listed in table 13 is contained in the following paragraphs.

In general, the materials in the coating of a shielded arc electrode determine the polarity and control the amount of penetration. E6010 electrodes give greater penetration than E6012 under identical conditions. This does not necessarily mean that greater penetration is obtained with reverse polarity. It merely means that coatings of the type used on the E6010 class usually result in electrodes that give greater penetration. Electrodes operating on either polarity usually melt at a higher rate when the electrode is negative (straight polarity). Therefore, higher travel speeds are ordinarily employed, resulting in less penetration. Ordinarily, knowing that a straight polarity electrode gives less penetration, is sufficient.

ALL-POSITION ELECTRODES

There are two subtypes under this group— reverse-polarity all-position electrodes and straight-polarity

ALL-POSITION ELECTRODES.

The reverse-polarity shielded arc electrodes have a paste coating of appreciable thickness which is comprised essentially of cellulosic materials (wood, flour, etc.). The straight-polarity, all-position electrodes also have an appreciable coating thickness, but the coating in this instance is made up of cellulosic and mineral materials. The lesser coating thickness, together with the arc action created by these

materials, makes it possible to weld in the flat, horizontal, vertical, and overhead positions with electrodes of diameters up to and including three-sixteenths inch.

E6010-All-Position, Direct Current, Reverse Polarity

The E6010-all-position, direct current, reverse polarity electrode is the best adapted of the shielded arc types for vertical and overhead welding. It is, therefore, the most extensively used electrode for the welding of steel structures which cannot be positioned and which require considerable welding in the vertical and overhead positions. The quality of the weld metal is of a high order and the specifications for this classification are correspondingly rigid. The essential operating characteristics of the electrode are:

1. Strong and penetrating arc, enabling penetration beyond the root of the butt or fillet joint.

2. Quickly solidifying weld metal, enabling the deposition of welds without excessive convexity and undercutting.

3. Low quantity of slag, with low melting and low density characteristics so as not to interfere or become entrapped when oscillating and whipping techniques are used.

4. Adequate gaseous atmosphere to protect molten metal during welding.

Electrodes of this type are usable only with direct current on reverse polarity (electrode positive).

E6011-All-Position, Alternating Current

The E6011-all-position, a. c electrode is intended to have similar operating characteristics and to be used for similar welding applications as E6010, but with alternating current.

E6012-All-Position, Direct Current, Straight Polarity

The E6012 type of electrode is often referred to as a "poor fit-up electrode" because of its ability to bridge wide gaps in

joints. It is particularly well adapted for single-layer welding of horizontal fillets.

This type is very extensively used in steel fabrication because it offers economy due to ease of use and high welding speeds. The usual operating characteristics of this type of electrode are listed as follows:

1. Suitable for use with direct current, with either straight (electrode negative) or reverse

(electrode positive) polarity, and with alternating current. Straight polarity is preferred because of a more direct and stable arc.

2. Adequate penetration in order to reach the root of fillet and other joints, but not as deep penetration as with E6010 in order to enable the filling of wide gaps without burning through. The small diameters, as three-thirty-seconds and one-eighth inch, are especially adapted for sheet metal welding without tendency to burn through the sheets.

3. The slag is more abundant and covers more of the pool than those of E6010 types, but it is not as abundant or as fluid as those of E6020 and E6030 types. The slag solidifies very rapidly just below the freezing point of the metal. The slag is generally dense and close fitting to the deposit.

4. The molten metal may be considered slightly more fluid than that of the E6010 type, but not to the extent that this electrode cannot be used in all-position welding. The molten metal and slag characteristics control the shape of the weld deposit and make the electrode especially suitable for horizontal fillet welding, producing flat or slightly convex beads without undercutting. Many of the electrodes of this type are suitable for vertical welding in the downward direction, although, for some purposes, the penetration and throat thickness are insufficient. In vertical welding in the upward direction, small welds are more convex and with wider spaced ripples than with E6010 electrodes.

E6013-All-Position, Alternating Current

The E6013 type of electrode is intended to be similar to E6012 type, but with improved arc characteristics with alternating current. These improvements are necessitated by the fact that many types of alternating-current machines have a relatively low open-circuit voltage and demand easily ionized coating materials in the electrode for satisfactory operation. The applications for this electrode are the same as for E6012.

FLAT-POSITION ELECTRODES

Electrodes of this type are covered by classifications E6020 and E6030. These electrodes usually have a very heavy mineral coating consisting principally of metallic oxides, asbestos, clay, or silicates. These electrodes depend upon the slag for the shielding action. They are used with high currents, and the molten metal is very fluid. This combination of extreme fluidity of the molten metal, together with the large amount of slag, makes it practically impossible to weld vertically and overhead with these electrodes.

E6020-Horizontal Fillets and Flat Positions, Direct or Alternating Currents

The E6020 electrode is designed for the production of flat or concave surface fillet welds in the flat or horizontal positions, with either direct or alternating current. This electrode has numerous applications where very high quality weld metal is required and where work is positioned, such as in the fabrication of pressure vessels, machine bases, gun mounts, and similar structures. The essential operating characteristics are as follows:

1. Usable with either alternating or direct current. It is mainly used with alternating current, but when used with direct current, straight polarity is preferred, especially for the welding of horizontal fillets.

2. The main requirement of the electrode is to produce horizontal fillets of flat or concave surface without undercutting. This necessitates that the molten metal and slag be comparatively fluid, that the metal be quick-freezing, and that the slag continuously cover the back portion of the pool and actually wet the molten metal.

3. Slags of this type are not quick-freezing, but remain as a plastic glass for some time

after the molten metal has solidified. The slag and metal are both too fluid to permit general welding in vertical or overhead positions.

4. While specifically designed to meet horizontal fillet welding requirements, the electrode is also adapted to the welding of butt or other flat position joints.

5. The physical properties of welds are of a very high order, especially in elongation. Radiographs show that properly made welds are practically perfect.

E6030-Flat Position, Direct or Alternating Currents

The E6030 type overlaps with E6020, but particular attention is given to its use in V and groove butt welds. It differs from E6020 in that it produces a smaller amount of slag and a less fluid slag, thus decreasing the possibility of slag interference and deep grooves. It is not as suitable for horizontal fillet welding due to insufficient slag coverage of the molten pool.

Electrodes of this type are used in flat position welding of pressure vessels and numerous other objects. It is necessary that the weld deposit meet the highest standards of X-ray and physical property requirements. Essential operating characteristics are listed as follows:

I. Adaptability for use in narrow- or wide-groove butt joints, providing adequate slag coverage for weld

shape and protection but not slag interference with the arc

2. Slag must wet metal surface and produce a concave weld within the confines of the groove.

3. The slags are porous and, therefore, are very easy to remove. They have a hardening range extending considerably below the freezing point of the steel deposit.

ELECTRODE CLASSIFICATION BY BEAD SHAPE

The Aviation Structural Mechanic will arc weld various types of steel more often than the nonferrous metals. Therefore, the following method of classifying electrodes for steel welding should prove very useful.

Classifying electrodes for steel welding is based on the shape of the bead which is obtained when arc welding with a particular electrode. These shape classifications are defined in the following pages.

Figure 60.—Flat bead produced by class E6010 •iMtrod*.

Flat-Bead Type (Class E6010)

The E6010 flat-bead type may be classified as a general purpose electrode because it is used for a wide variety of work and possesses high average physical characteristics. It

has a heavy coating and is best suited for direct current with electrode positive.

In smaller sizes—% 2 -inch and smaller—the E6010 is suitable for use in all positions. The % 6 -inch size is also made special for all positions. It is suitable for fillets, deep grooves, and all types of joints in all sizes. It has deep penetration qualities and is used very satisfactorily on square butt joints where the electrodes actually scarf or melt the plates. It produces a rather flat bead of general type as shown in figure 60.

Convex Bead Type (Class E6012)

The class E6012 convex bead type of electrode has a heavy covering and gives best results with direct current with the electrode negative, or it may be used satisfactorily with alternating current.

I ^ m [

Figure 61.—Smooth, convex bead produced by class E6012 electrode.

Sizes %2-inch and smaller are suitable for all positions— although % 6 -inch often is used in all positions—and in larger sizes for welding in flat positions. It may be used for fillet welding, single or multiple pass, and can be used for butt welds of the V- or U-groove type.

Because of its deposition characteristics, and ability to build-up, it is frequently used where fit-up is poor or where

a small admixture of base metal is desired. It produces an exceptionally smooth bead which is somewhat convex, as indicated in figure 61.

Conceive Bead Type (Class E6020—-E6030)

The concave bead type of electrode has a heavy covering and can be used with direct current and with the electrode either positive or negative, or it can be used with alternating current. It is used in the flat position only, and is not suitable for vertical or overhead work.

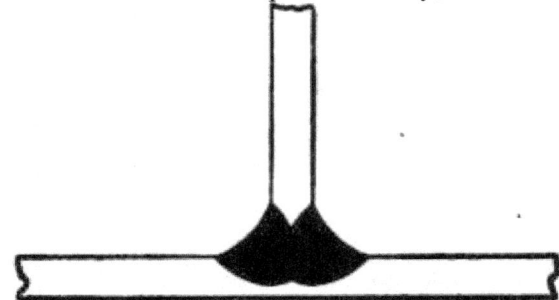

H0W* 62.—Smooth, concovo bood producod by cIosmi E6O20 and E6030.

This type is used for fillets or butt joints of the V- or U-groove type. It flows very readily, producing a heavy slag cover on the weld. It is sometimes known as the "hot rod" type. It produces a very smooth bead, slightly concave, as shown in figure 62, when welded in grooves or position fillets, but not so concave when fillets are horizontal.

HOW THE ARC WORKS

Electric arc welding, as we have seen, is a fusion welding process based on the principle of generating heat with an electric arc jumping an air gap to complete an electric circuit. This process develops considerably more heat than an oxyacetylene flame. In some applications it reaches a temperature of approximately 10,000° F.

In the basic electric arc welding process, electricity is generated by either a motor generator welding machine or stepped up by a transformer.

The motor generator or the transformer, whichever is used, has two terminals to conduct the electricity. A metallic wire electrode held in a suitable holder is connected to one terminal, and the metal to be welded to the other terminal. When the electrode is touched to the metal being welded, the electrical circuit is completed and the current flows. When the electrode is withdrawn from the metal, an air gap is formed between the metal and the electrode. If this air gap is of the proper length, the electric current will bridge this gap to form a sustained electric spark called the electric arc.

The instant the arc is formed, the temperature of the work at the point of welding and the welding electrode increases to approximately 6,500° F.

INSULATED CABLE

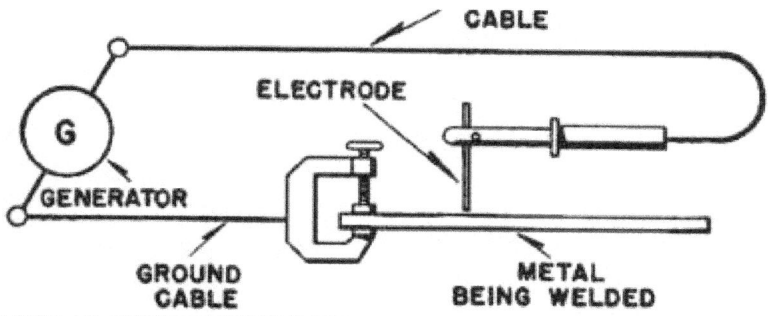

METAL BEING WELDED

Figure 63.—Arc welding circuit.

This tremendous heat is concentrated at the point of welding and in the end of the electrode, and simultaneously melts the end of the electrode and a small part of the work to form a small pool of metal commonly called the crater. Under the intense heat developed by the arc, a small part of the work to be welded is brought to the welding point almost instantaneously, and tiny particles of molten metal are formed at the end of the electrode. These tiny particles, or globules, are

then forced across the arc and deposited in the molten crater in the work. Because of this fact it is possible to make overhead welds.

The use of electric arc welding in aircraft construction is a comparatively recent development, and is rapidly gaining in popularity as the process is continually improved. Electric arc welding is currently being used to fabricate aircraft engine exhaust collectors, exhaust manifolds, engine mounts, landing gear parts, gasoline and oil tanks, and so forth.

An increasing number of aircraft parts are being manufactured by the arc welding process, and these same parts also may be repaired by this method. It can be expected, as improvements continue, that the Aviation Structural Mechanic will use this process more and more in aircraft maintenance and repair. However, the principal uses of electric arc welding at the moment are in shop maintenance and repair. Such jobs include fabricating and maintaining jigs, tools and fixtures; replacing iron castings with lighter, stronger, and less costly steel; eliminating bolts, rivets, screws, and other less efficient methods of joining metals; repairing broken machinery and rebuilding surfaces which have been worn or damaged in use; filling sand holes or other imperfections in castings; and reclaiming improperly machined parts.

Before attempting to weld, something about the equipment used for arc welding must be considered. In our discussion, each necessary piece of arc welding equipment will be described. Included in the information will be a general description of the piece of equipment, its use and operation. When information is desired regarding the operation or particular use of any specific piece of equipment, refer to the manufacturer's instructions which usually accompany it This applies particularly to arc welding machines.

WELDING MACHINES

The principal piece of arc welding equipment is the machine which supplies the electricity. Arc welding

machines are classed as either direct or alternating current units, depending upon the type of current they deliver to make the arc. Each has advantages and disadvantages peculiar to the type of current produced.

Advantages such as initial cost, portability, and operating expenses are claimed for each class of welding machines. However, from a welder's standpoint, d. c. equipment offers fine current adjustments and choice of polarity with its attendant advantages. An a. c. welder eliminates practically all arc blow which is troublesome to welders using d. c. equipment, especially when making heavy fillet or deep groove welds.

A direct-current welder (fig. 64) consists of a direct-current generator driven by a suitable type of motive power. The voltage of such a generator will usually range from 15 to 45 volts across the arc, although any setting is subjected to constant variation due to arc conditions.

A fairly wide range of current output is necessary to accommodate the various kinds of work. The range of current will vary also, depending upon the type of unit and work for which it was designed.

In direct-current welders, the generator is of a variable voltage type and so arranged that the voltage automatically adjusts itself to the demands of the arc. However, the open-circuit voltage is manually set to the correct range by means of a rheostat mounted on the control panel. Amperage of the welding current is manually adjustable, and is usually set to the proper range by means of a reactance arrangement, or a selector switch which taps into the field coils of the generator at different points to increase or decrease its strength. When both voltage and amperage are adjustable by means of individual controls, the machine is referred to as a dual-control type. The dual-control machine is the most popular in use today.

Another system employed to a limited extent makes use of adjustable generator brushes for the control of the current. Machines of this type are provided with one control which proportionately varies both amperage and voltage by the movement of the brush assembly. .

When a power supply is available, welding generators are driven by means of an electric motor. The armatures of both generator and motor are usually on a single common shaft. However, some of the older machines are arranged whereby the motor drives the generator through a flexible coupling.

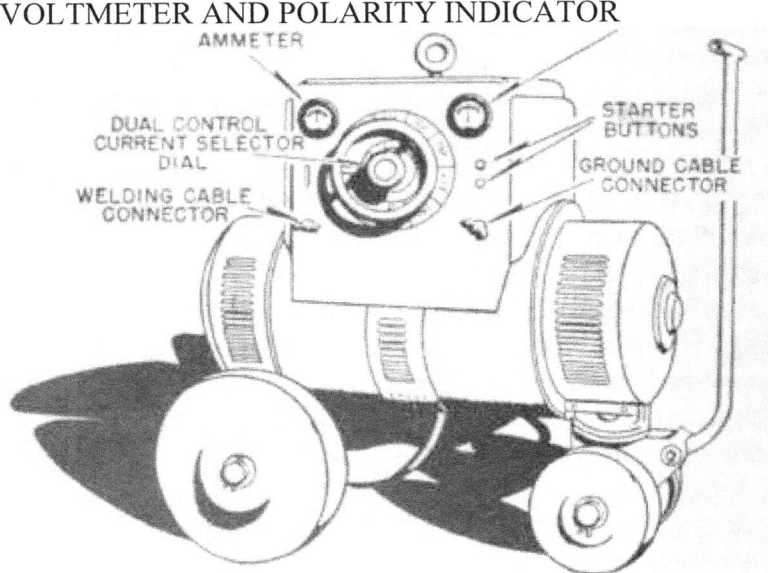

Figure 64.—Direct current welding machine (dual control).

In many instances, arc welding must be performed in places where a power supply is not available. Portable gasoline engine-driven generators are particularly adaptable in such cases, as they are available in compact units easily transported from place to place. The engine used for this purpose must be fitted with a suitable governor to compensate for the varying loads imposed by the welder.

Alternating current welding machines are divided into two general classifications—the transformer type and the motor-generator type. The latter type is so infrequently seen that it is not included in this discussion.

The transformer type derives its welding current from a core transformer. The primary coil is connected directly to the power line, and the secondary coil supplies the welding current. Current control, which varies in different models, is accomplished by either a bridge reactor, movable coil, movable core, or by tapping the secondary coil. All means of current control, except the tapped secondary method, offer continuous control.

CURRENT SELECTOR

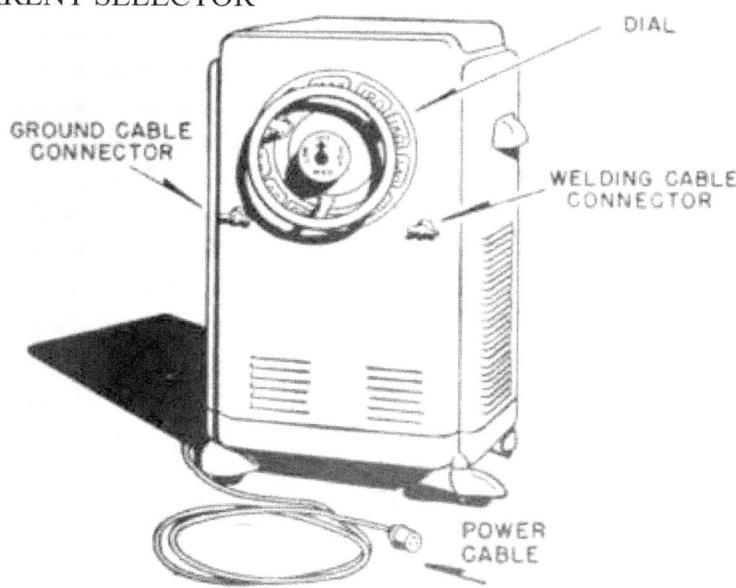

Figure 65.—A. c. transformer type welding machine.

The transformer type a. c. welder (fig. 65) is rapidly gaining in popularity because of advances in machine design and the development of heavy-coated electrodes especially designed for alternating current welding.

Electric arc welders are manufactured in sizes ranging from 100 to 600 amperes for manual welding. Machine sizes are based on their amperage output. For example, a 100-ampere machine will deliver 100 or more amperes (the output is rated conservatively by most manufacturers). Naturally, the range in sizes of arc-welding machines is

238500° .-»:: 13

governed by the class and range of work for which they are to be used.

The 200-ampere machine is the machine which most economically satisfies the needs of aircraft welding. A machine of this size will rapidly weld the light metals used in aircraft and can also be used for most of the maintenance in the shop. For such work, as found in bridge construction and other heavy fabrication, a 300- to 600-ampere machine is used.

Maintenance

Because of the amount of dust and grit present in all welding shops, the problem of proper maintenance is a very important one. The following instructions will apply to most arc-welding machines.

Forced draft is used to cool most welding machines, and because of this fact particles of dirt are carried throughout the unit. Under average conditions, the machine should be cleaned with dry, compressed air at least once each month. This may readily be done by removing the dust covers and shields. Should the machine appear greasy at the time of cleaning, it should be dismantled and thoroughly washed with carbon tetrachloride. During the regular monthly cleaning, an inspection should be made of the condition of the switch points, brushes, commutator, and bearings.

The machine should be given a thorough greasing at 4- to 6-month intervals. This may follow the cleaning operation and should include all bearings in the unit. Too much grease may be as harmful as not enough, as a surplus may be thrown upon the commutator. Grease on the commutator may result in serious damage or a fire hazard.

To properly grease a bearing of the type used in welding machines, the plug should be removed on the lower side of the bearing boss and the machine started. Grease may then be injected into the fitting until it begins to emerge from the plug hole. Allow the machine to run for several minutes to force out any pressure on the grease, then replace the plug. Only an approved grease should be used.

The brushes and commutators of both the motor and generator are subject to considerable wear. Brushes worn so that spring tension is appreciably reduced must be replaced to maintain proper efficiency of the machine,. Although new brushes are formed to fit the commutator, they must often be sanded in to give proper amount of contact. This is done by wrapping a strip of No. 00 sandpaper around the armature and turning it by hand until the brushes have been worked down to a perfect fit. Never use emery cloth.

Brush springs weakened from overheating should also be replaced to assure positive brush contact.

Each time brushes are replaced, the commutator should be checked for cleanliness and wear. If a deposit of graphite from the brushes is found, it may be removed by holding a piece of No. 00 sandpaper against the commutator while the armature is in motion. Ridges or pockets on the surface of the commutator will require the removal of the armature so that it may be "trued up" on a lathe. Only a light cut should be taken, and the mica separators between the bars of the commutators must be undercut from one sixty-fourth to one thirty-second inch after the truing operation. Although a special cutter should be used for this purpose, a hacksaw blade will serve in an emergency.

All electrical switch contacts should be sanded clean if pitted. Parts that have been badly burned should be replaced. At least once each year, the windings of the generator and motor should be inspected and, if found dry or cracked, coated with shellac.

WELDING CABLE

The welding current is conducted from the generator to the work by multistrand, well-insulated copper cables, two of which are required to complete the circuit between the welding machine and the work. An extra flexible cable is used between the welding machine and the electrode holder. The ground cable, which is connected between the work and the machine, need not be as flexible as the electrode holder

cable, although similar types of cable are sometimes used for both.

Flexible cable is designed especially for welding. It derives its flexibility from its construction, since it is made of thousands of very fine wires enclosed in a durable paper wrapping which allows the conductor to slip readily within its rubber insulation when the cable is bent The rubber also contributes to its flexibility. The ability of the cables to withstand wear and abrasion is provided by a tough, braided cotton reinforcing and by the composition and curing of the waterproof rubber covering, which also provides a smooth finish.

The size of the cable is determined by the size of the welding machine and the distance of the work from the machine. As these factors increase, the size of the welding cables must also increase. When the cable is too small in relation to the amperage used, it will become overheated. A cable which is too small will not carry sufficient current to the arc without overheating, but the larger the cable the more difficult it is to handle.

Selection of the size of cable has a definite and important bearing on welding efficiency. Table 14 offers a guide for selection of the correct cable to use on various machines.

It is further recommended that the longest length of 4/0 cable for a 400-ampere welder should not be greater than 150 feet, and for a 600-ampere welder not more than 100 feet. For greater distances the cable sizes should naturally be increased despite the fact that cables of such length and size are difficult to handle. Rather than increase the size of the welding cable, move the machine closer to the work.

ELECTRODE HOLDERS

An electrode holder is essentially a clamping device for holding the electrode. It is provided with a hollow, insulated handle through which the welding cable is passed to connect with the electrode clamping device.

TabU 14 RECOMMENDED CABLE SIZES

The advantage of an insulated electrode holder lies in the fact that it may be touched to any part of the work without danger of short circuiting. The clamping device is made of an alloy which is a good conductor of electricity and durable under high temperature and constant use. The clamp is designed to hold the electrode securely in any position and to permit quick and easy change of electrodes. An electrode holder should be light in weight to permit ease of handling yet sturdy enough to withstand rough usage. Typical electrode holders are shown in figure 66.

Electrode holders are made in a variety of sizes, and each manufacturer has his own system of designation. However, any catalog description of an electrode holder gives the maximum amperage and range of electrodes by diameters which the holder will accommodate. The size used is dependent upon the amperage rating of the welding machine, that is, a larger holder must be used with a 300-ampere welder than with a 100-ampere machine. If the electrode holder is smaller than the type which should be used for a particular machine, the holder will overheat.

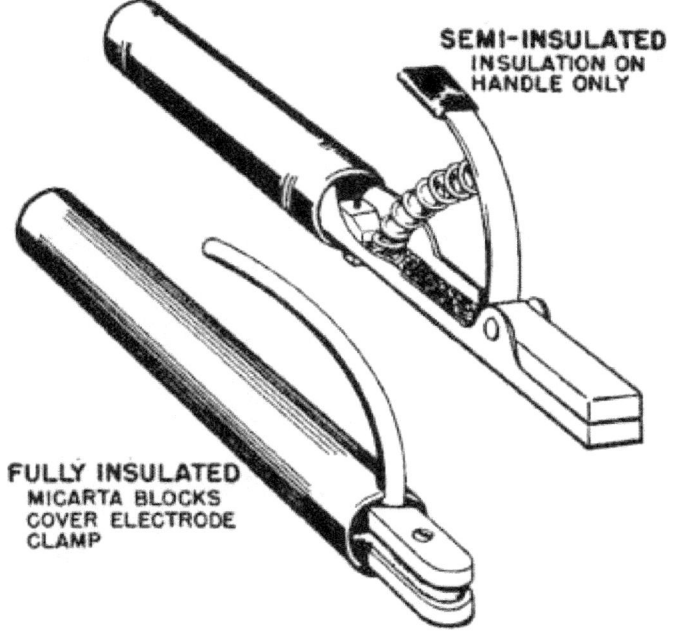

SEMI-INSULATED
INSULATION ON
HANDLE ONLY

FULLY INSULATED
MICARTA BLOCKS
COVER ELECTRODE
CLAMP

Figure 66.—Electrode holders.

PROTECTIVE EQUIPMENT

Exposure of the eyes to the infrared and ultraviolet rays accompanying electric arcs often results in eye-burn or "sand in the eyes," as it is commonly called by arc welders, and may cause extreme pain for 24 to 48 hours. In addition to affecting the eyes, exposure to the arc will produce severe sunburn to portions of exposed skin.

To protect the eyes and face, a head shield should be worn. These shields are constructed of pressed fiber and are solid black in color to reduce reflection. They are light in weight and designed to insure comfort to the welder.

Protective shields are provided with a glass window whose standard size is 2 by 4% inches. The composition of the glass is such that it absorbs the infrared and ultraviolet rays and most of the visible rays from the arc.

The welding lens is protected from metal spatter and breakage by a plain or a chemically treated, clear, non-

TobU 15.—RECOMMENDED WELDING GLASS NUMBERS
Welding
glass number
Arc welding appli-tion (amperes)
To 30
To 75
75 to 200. 200 to 400 Over 400.

spatter glass covering the exposed side of the lens. Green-tinted flash goggles should be worn by the welder to protect his eyes from the flashes of other welders when he removes the shield for purposes such as inspecting his weld. They also protect his eyes when he is grinding, chipping, or cleaning slag. Flash goggles should be worn by all persons working in the vicinity of arc welding operations.

During any arc welding operation, a continuous shower of sparks and hot molten metal are thrown off by the arc. These may cause burns if permitted to contact the skin; therefore, protective clothing must be worn to shield the welder from the spray and sunburn effects of the arc. Leather gloves, preferably of the gauntlet type, protective sleeves of leather, and leather aprons should also be worn at all times. Leather jackets provide excellent protection when welding in the vertical or overhead position. There are many other types of protective clothing, usually made of leather, such as trousers and leggings. Naturally, the type and amount of protective clothing you need will depend upon the position and size of the welding job.

In addition to wearing protective clothing, the welder should keep the collar button on his shirt fastened, and wear ankle-high shoes in preference to oxfords. It is also important that trousers without cuffs be worn and that the legs of the trousers be held down over the shoe tops. Bicycle

clips are excellent for this purpose. The welder can protect the top of his head with a brimless white hat. Woolen clothing which offers more protection than cotton, should be worn, if possible.

THE WELDERS HAND TOOLS

The tools shown in figure 67 are a necessary part of every welder s set of tools.

The steel wire brush and chipping hammer are used to clean the work before welding. Slag must be removed by chipping or wire brushing after the weld is completed and between each pass in multiple-pass welding. In addition to the chipping hammer and wire brush, an arc

welder has a definite need for the following listed hand tools:
1. Center punch.
2. Flat cold chisel.
3. Combination square
set.
4. Scriber.
5. Flexible steel rule.
6. Combination pliers.
7. Ball peen hammers.
8. Soapstone.

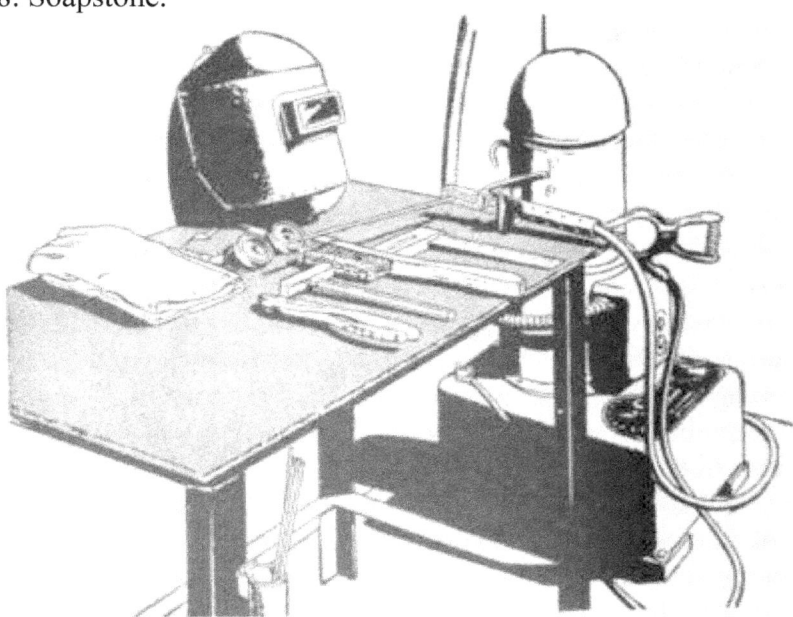

Figuro 67.—Are woldor's hand tools. 190

The following troubleshooting chart should be of value in determining common troubles encountered with arc welding equipment. The first column describes the trouble or the action of the equipment, and the second column gives the probable cause. The last column indicates the corrective measures to take after the cause of the trouble has been determined.

Tdbl* 16.—TROUBLESHOOTING CHART FOR ARC WELDING

Trouble

Machine fails to hold the "heat" properly.

Probable cause

Rough or dirty commutator.

Brushes may be worn down to limit of adjustment or life.

Brush springs may have lost adjustment or may be broken.

Field circuit may have variable resistance connection or intermittent open-circuit, due to loose connection or broken wire.

Electrode lead or work lead connections may be poor.

Wrong grade of brushes may have been installed on generator.

Field rheostat may be making poor contact and overheating.

Brush-shifting or other mechanical current-adjusting mechanism may have loose or worn links.

Current control brush holder contact springs may be worn out or
Remedy
Commutator should be trued or cleaned.
Replace or readjust brushes.
Replace or read just-brush springs.
Check field current with ammeter to discover varying current. This applies to both main generator and exciter if used.
Tighten all connections.
Check with manufacturer's recommendations.
Inspect rheostat and clean and adjust finger tension on switch.
Check current-adjusting mechanisms for backlash and play.
Inspect, replace needed parts, clean internal contact surface of con-
Trouble
Welder starts but fails to generate current.

Probable cause

bent. Contact surface may be dirty, rough, or pitted. Current control brush holder support stud and mating contact surfaces may be dirty or pitted and bumed.
Engine regulator shorting switch out of adjustment.
May be running the wrong way.
Generator or exciter brushes may be loose or missing.
Exciter may not be operating.
Field circuit of generator or exciter may be open.
Remedy
trol device. Do not lubricate. Smooth up roughened surface.
Clean off brush holder stud and internal contact surface—use light application of vaseline to stud and replace. If brush holder internal contact surface is burned replace brush holder and support stud.
Adjust switch contacts or mercury switch tilt angle so circuit is open when engine is at full speed and when welding.
Check direction of rotation with manufacturer's instructions or direction arrow. On 3-phase motors, direction of rotation may be changed by interchanging any 2 input leads.
Be sure that all brushes bear on the commutator and have proper spring tension.
Check exciter output voltage with voltmeter or lamp.
Check for open circuits in rheostat, field leads, field coils. Also check resistors and rectifiers if any. Some machines give less
Trouble
Probable cause
Remedy
Welding arc is loud and spatters excessively.
Welding current too great or too small compared to indication on the dial.
Generator may be reversed in polarity due to another machine or incorrect operation in parallel with another machine.
Series field and armature circuit may be open-circuited.
Reversing switch wiper contact bent and not clearing the blade of switch when switch is closed.

Current setting may be too high.

Polarity may be wrong.

Engine regulator shorting switch closes intermittently when running at full speed, causes increasing surge of current and spatter.

Current control, shaft and handle may have turned slightly in the insulated bushing of the current control brush holder, caused by turning handle

output when fields are ooen.

Flash the field with a storage battery or another generator, first with one polarity and then with another to see if it "builds up" (flash exciter field,if set has separate exciter).

Check circuit with ringer or voltmeter.

This shorts the exciter series and causes failure to generate—bend or replace to secure correct operation.

Check setting and current output with ammeter.

Check polarity. Try reversing polarity or try an electrode of the opposite polarity.

Adjust so contacts are well open or mercury level well below contact on mercury tilt switch when engine is at full speed position of engine regulator (bellows fully extended).

See that Current Control indicator yellow arrow is in the horizontal position when handle is turned against stop in the minimum direction.

Trouble

Probable cause

Remedy

too hard against one of the stops. Exciter output low causing low output compared to dial indication.

Current Control set to minimum and welder output so great that motor stalls when arc is struck.

Motor trips off the line.

Power circuit may be single phased.

Welder may be operating above current capacity.

Welding electrode or work leads may be too long or too small in

Surrounding atmospheric temperature may be too high.

Motor input voltage too low (or too high) under load.

Field discharge resistor wired to reversing switch and open-circuited. Check for circuit through it.

Motor is probably running backwards or series fields connected reversed to make a cumulative series generator. Check rotation.

Check for one blown fuse or dead line.

Check load against welder nameplate.

Check terminal voltage while machine is loaded; it should not exceed 40 volts when operating at rated current.

Make sure that temperature in motor-generator room or housing does not exceed 100° F., and that there is no interference with normal ventilation of the machine.

Motor supply voltages should not fall below 90 percent of normal voltage. Have power company check transformer and line capacity. If supply

Trouble

Probable cause

Remedy

leads too long or too small should be corrected.

Machine fails to start.

Power circuit may be completely dead.

Power circuit may be single phased.

Power-line voltage may not be suitable for motor, or may be extremely low; may be accompanied by chattering of the motor starter.

Machine may be jammed.

Motor starter may be single phased.

Overload protecting device may be tripped or contacts open-circuited.

Look for open disconnect switch, fuses removed from clips, or blown fuses.

Look for one blown fuse or one dead line.

Check voltage with voltmeter, particularly at the moment of attempted starting.

See that armature turns over easily by hand, and look for foreign matter in air gaps

Check to see that all fingers on starter make contact when closed.

If machine has had time to cool after tripping due to overload or is cold and starter fails to close, check for circuit through push button, N. V. R. coil and thermostats to find the open-circuited part. See wiring diagram for normally closed and open contacts on the push button.

QUIZ

1. What is arc force?
2. To what is the degree of penetration proportional?
3. What determines polarity ?
4. How may polarity be determined?
5. From what does arc blow result?
6. Name the different types of electrodes.
7. Classify electrodes by bead shape.
8. How is electricity obtained for the arc welding process?
9. What is the dual control type of welding machine?
10. How often should the welding machine be greased?
11. What is an electrode holder?
12. Why should a head shield be worn when you are welding?
13. Why is leather used for protective clothing?
14. What are the two most important hand tools a welder uses?

CHAPTER 10

ARC WELDING PROCESSES AND TECHNIQUES PRELIMINARY TECHNIQUES

As a preliminary preparation to welding, it is important to ascertain that welding equipment is in good working order and that all connections are correct and tight. Particular attention must be paid the ground, as a poor connection will result in a fluctuating arc difficult to control. To make good contact, the clamp should be clean and the point of contact thoroughly brushed.

The electrode is clamped to its holder at right angles to the jaws. Shielded electrodes have an end of the electrode

free of coating to provide good electrical contact. Handle the electrode holder with care to prevent accidental contact with the bench or work, as such contact may weld it fast.

WELDING MACHINES

The machine most likely to be found in the welding shop will probably be a d. c. dual control type, rated at 200 amperes, and therefore this type of machine will be used in our discussion. As previously stated, these units have two controls^—one for open-circuit voltage, the other for amperage.

Manufacturers of welding machines use different dial markings, and it is advisable to refer to the manufacturer's operating manual for specific information concerning a particular machine. This manual will furnish the necessary information for adjustment to proper amperage and voltage values necessary for particular jobs. The general procedure, regardless of the machine dial markings, will follow the pattern described below.

After checking the polarity of the machine to make certain that it coincides with the electrode used, set the machine at the highest open circuit voltage within the limits of the job. This makes it easier to maintain the arc. Then set the amperes at the lowest recommended ampere setting, according to the diameter of the electrode. This is the preliminary setting.

Begin the weld at this setting and increase the amperes (heat) and lower the voltage until the arc and the resulting weld are satisfactory. Most welders run an experimental weld until the machine is adjusted to their individual liking. t Good welders weld as much by ear as they do by eye, and they judge a good arc by the satisfying frying sound which it makes.

There are a number of variable factors affecting the machine setting. These include size and type of electrode, thickness of metal to be welded, type of joint, and skill and technique of the welder. With these variables to be considered, it is apparent that any set of current values could be merely

generalization. Current values as published by different manufacturers vary considerably for the same classification and size of electrode.

Table 17 is included for your information, but the current values in this chart are merely suggestive. A setting on the welding machine within these ranges should be used only as a preliminary setting since the table is intended to cover all welding positions.

The proper welding current for a given set of conditions can be determined from the degree of electrode heat. If

Tobte 17.—RECOMMENDED AMPERAGE AND OPEN CIRCUIT VOLTAGE RANGES
238500° -14 1M

the electrode becomes excessively hot, it indicates that the current is too high. Welds of good quality cannot be made if the electrode overheats, and in such instances the current must be reduced or the size of the electrode increased. With proper current and electrode, a smooth, uniform bead should result.

Relaxation while welding is important. Gripping the electrode holder too tightly causes the muscles used to control the electrode to be under tension, and the welder tires easily and loses the control essential to a good weld. To

Figure 68.—Proper technique for holding arc welding electrode, showing the two-handed grip.

relieve some of the electrode holder's weight, the cable may be draped over the welder's shoulder or coiled in his lap. The holder is usually gripped in one hand which may be supported by the other. You should use the position which is most natural and which suits the position of the job being performed.

Most of the exercises in the appendix of this book are intended for performance in a sitting position. The hand gripping the electrode holder (see fig. 68) is supported by the other hand for added steadiness. Elbows are kept close to the body and the cable is draped over the shoulder—an excellent position for the beginner. When you have learned to control the electrode with both hands and have gained confidence, you should develop the ability to control the electrode with one hand.

Before striking the arc, the following list of items should be checked off.

h Is the machine in good working order?

2. Have all connections been properly made? Will the ground connection make good contact ?

3. Has the proper type and size of electrode been selected for the job?

4. Is the electrode properly secured in the holder?

5. Has sufficient protective clothing been provided, and is it in good condition ?

6. Is the work metal clean?

7. Does the polarity of the machine coincide with that of the electrode?

8. Is the machine adjusted to provide the necessary current for striking the arc?

When you have the correct answer to these questions, you are ready to begin arc welding.
STRIKING THE ARC

The welding arc is established by touching the plate with the electrode and immediately withdrawing it a short

distance. At the instant the electrode touches the plate, a rush of current flows through the point of contact. As the electrode is withdrawn, an electric arc is formed, melting a spot on the base metal and the end of the electrode.

The main difficulty confronting a beginner in striking the arc is freezing—that is, sticking or welding the electrode to the work. If the electrode is not withdrawn promptly upon contact with the plate, the high amperage will flow through the electrode and practically short circuit the welding machine. The heavy current melts the electrode which sticks to the plate before it can be withdrawn.

There are two essentially similar methods of striking the arc The first is a vertical up-and-down tapping motion, illustrated in figure 69. While this method is commonly

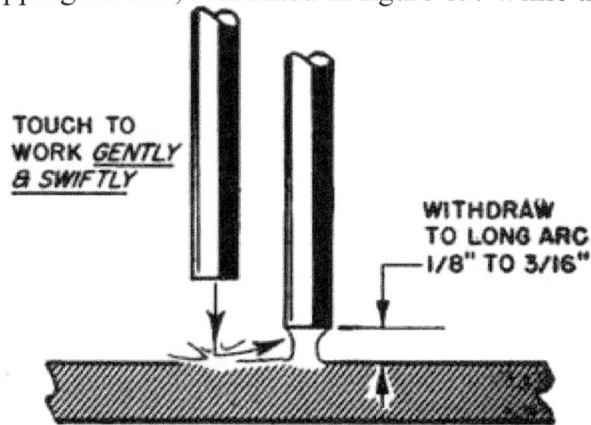

Figure 69.—Touch method of starting th« ore.

used by experienced operators, it often presents difficulties to the beginner. The second method of striking the arc, shown in figure 70, consists of a side-scratching motion of the end of the electrode in which the electrode tip barely grazes the surfaces of the plate making contact and establishing the arc.

Regardless of the method used, the electrode must be withdrawn quickly upon contact with the plate so as to provide the gap necessary to maintain the arc.

Try the first, or touch method for striking the arc, as shown in figure 69. Hold the electrode in a vertical position, lowering it until it is an inch or so above the point where the arc is to be struck. Hold it in this position without touching the electrode to the plate, and lower the face shield into position. Touch the electrode very gently and swiftly to the work, using a downward motion of the wrist, and immediately withdraw it to form a long arc (one-eighth inch to three-sixteenths inch). Hold the arc for a few seconds, then break it.

To strike the arc by the scratch method, move the electrode downward until it is just above the plate and at an angle of 20° to 25°, as shown in figure 70. Hold it there without

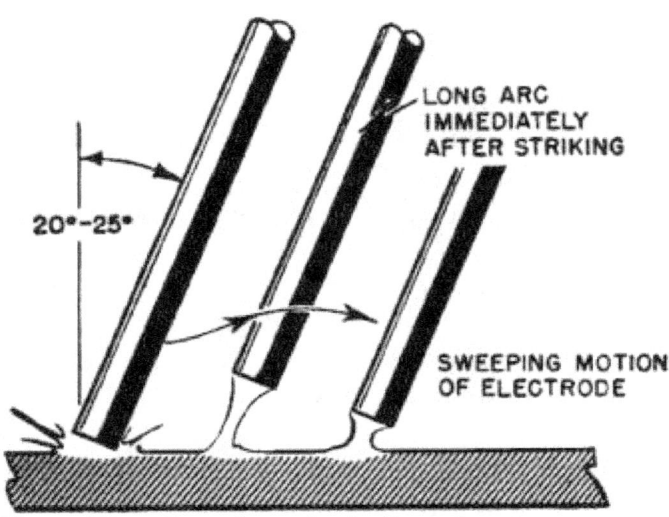

Figure 70.—Scratch method of starting tho arc.

touching the plate, then drop the shield to protect the eyes. Strike the arc gently with a swiftly sweeping motion, scratching the electrode on the work with a wrist motion. Immediately withdraw the electrode to form a long arc. Hold the arc for several seconds, then break it.

The purpose of holding an excessively long arc immediately after striking is to prevent the large drops of metal,

passing across the arc at this time, from shorting out the arc and thus causing freezing. This also helps to more smoothly fuse one bead with the previously deposited bead Practice striking the arc until proficiency and skill are attained. After the arc can be struck with ease, hold it long enough to run a bead of about half an inch. Remember that a good arc, with correct current value and length, is characterized by an unmistakable frying sound.

LENGTH OF WELDING ARC

The distance through the center of the arc from the electrode end to the point where the arc contacts the metal is referred to as arc length. With coated electrodes, the length is measured from the metallic core rather than the coating because the metallic core may burn away more rapidly than the coating.

Bare electrodes generally use an arc length equal to their diameter. Too long an arc results in poor fusion, excessive spatter, and a contaminated weld. Too short an arc may cause a very porous weld and may include particles of slag. In practice, the arc length will be determined by the kind of electrode, its diameter, position of welding, and amount of current used.

RUNNING A BEAD

A weld is a single bead or a combination of beads, and it is thus important for the arc welder to understand the difference between good and bad beads.

A bead is the metal deposited by one pass of the arc welding electrode. When a weld is made up of more than one bead, it is called a multiple pass weld.

To form a uniform bead, the electrode must be moved along the plate at a constant speed, in addition to the downward feed of the electrode. The rate of advance, if too slow, will form a wide bead resulting in overlapping, with no fusion at the edges. If the rate of advance is too fast, the bead will be too narrow and have little or no fusion at the

plate. When proper advance is made, no overlapping occurs, and good fusion is assured.

In advancing the electrode, it should be held at an angle of about 20°-25° in the direction of travel, as may be seen in figure 72.

RESTARTING THE ARC

If the arc is broken during the welding of a bead, a crater will be formed at the point where the arc ends. The arc may be broken by feeding the electrode too slowly or too fast, or when the electrode should be replaced. The arc should not be restarted in the crater of the interrupted bead, but just ahead of the crater on the work metal. Then the electrode should be returned to the back edge of the crater.

From this point, the weld may be continued by welding right through the crater and down the line of weld, as originally planned. (See fig. 71.)

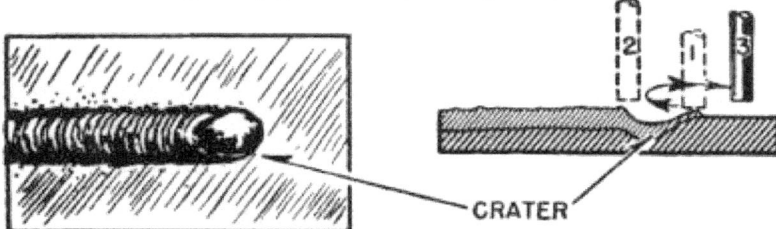

Figure 71.—Restarting the arc.

Every particle of slag must be removed from the vicinity of the crater before restarting the arc. This prevents the slag from becoming trapped in the weld.

After learning to weld straight line beads, it is good practice to weld in one direction to the end of the plate, move slowly to the side, then reverse the direction of travel.

Spend sufficient time on this exercise to become proficient in holding the proper length of arc and moving the electrode

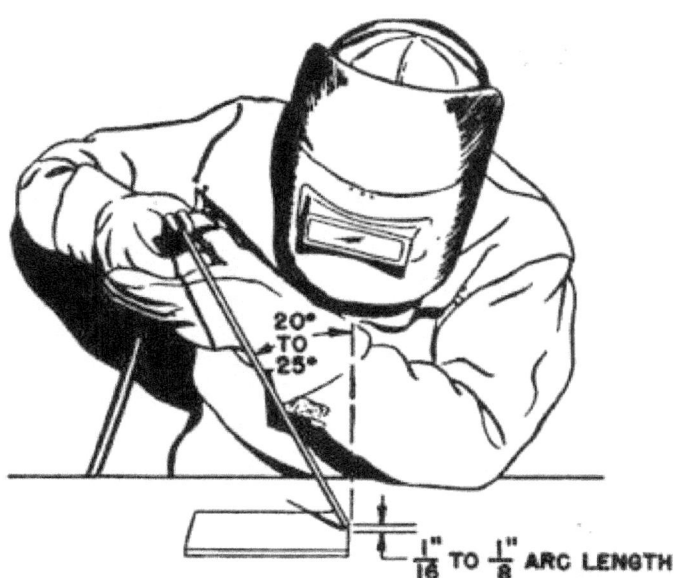

Figure 72.—Angle of electrode.

along the plate at the correct speed, so as to secure smooth, even beads.

Study the accompanying chart in table 18. Then continue practicing until you are able to make a weld by the correct procedure indicated.

WEAVING TECHNIQUE

When depositing weld metal, it is often desirable to make the width of the deposit wider than is obtained by depositing a single bead. This is accomplished by a technique known as weaving, or moving the electrode from side to side during the forward motion.

There are a number of different weaving motions used in welding, but in all cases it is important that the motion used be uniform. Typical weaving motions are illustrated in figure 73.

If the weave used is not uniform, or close enough, there is danger of poor fusion at the edges and of slag being trapped in the center.

PREHEAT "APPLY"

loom mm

HALF-MOON

SPREAD HALF-MOON

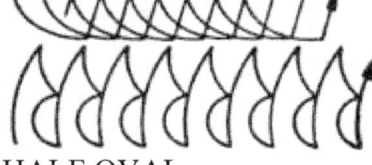

HALF OVAL

INVERTEO "F M

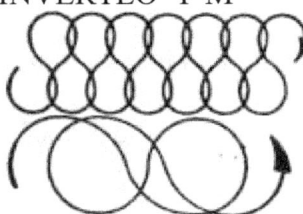

FIGURE M 8" BUTTERFLY rigure 73.—Typical weaving motions.

MULTIPLE PASS WELDING

Groove and fillet welds in heavy metals often require the deposit of a number of beads in order to complete a weld. It is important that the beads be deposited in a predetermined sequence in order to produce the soundest welds with the best proportions. The number of beads is, of course, determined by the thickness of the metal being welded.

The sequence of the bead deposits is determined by the kind of joint and the position of the metal. All slag must be removed from each bead before another bead is deposited. This is most important, and will be discussed more fully as our study progresses.

PREPARATION FOR ARC WELDING

The strength of any weld may be appreciably affected by lack of proper preparation of the work. Better strength

is always obtainable when the work metal is clean and free of foreign matter. It is also highly important that the edges be prepared in a manner that will permit complete fusion without an excessive amount of heat. This is also necessary in order to minimize the amount of heat radiating from the weld to the surrounding base metal

The same five types of joints are used to weld various forms of metal by either the oxyacetylene or the electric arc method. These are the butt joint, tee joint, lap joint, corner joint, and the edge joint. The kind of joint, thickness of metal, direction of welding, facilities for preparing the metal, and the load to which the weld is to be subjected govern the preparation of the joint. (See fig. 74.)

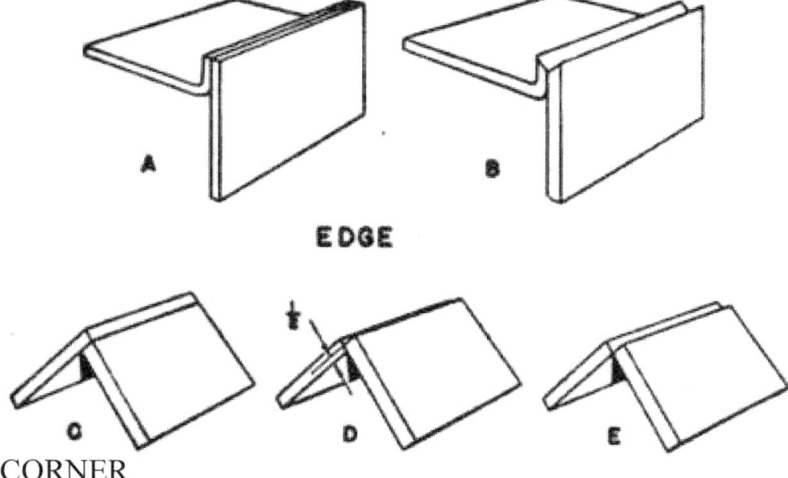

EDGE

CORNER

Figure 74.—Preparation of edge and corner joints for arc welding. A For metals 18 to 12 gage, B for 10-gage metals, C for heavier sheet.

Welds are commonly identified by the kind of joint involved, being referred to as butt welds, lap welds, edge welds, tee welds and corner welds. Obviously, a so-called butt weld may be of either the bead or groove type, according to the preparation of the joint.

SPECIFICATIONS FOR ARC WELDS

The specifications which follow are for arc welding required in general maintenance and repair work, such as the manufacture of work stands, storage racks, etc. When welding aircraft parts, approximately the same specifications apply to arc welding as to oxyacetylene welding of these parts.

Bead Weld Specifications

A bead deposited on a metal surface to build it up to a greater height or thickness should be approximately 1% times as wide as the diameter of the electrode being used. Generally, an inch of bead length should be deposited for every inch of electrode used. Beads adjoining other beads on the base metal should be fused to one-fifth of their width on either or both sides, depending on the situation.

Groove Weld Specifications

The depth of the throat designates the size of a groove weld, whether it be in a butt joint or an outside corner joint. If the plates being welded are of different thicknesses, the thickness of the lighter plate designates the size of the weld. The amount of metal extending above the surface of the base metal is called reinforcement, and it may range from one-thirty-second to one-eighth inch. If the amount of reinforcement is greater than one-eighth inch, the joint will probably concentrate stresses at the edge of the groove, rather than increasing the strength of the joint. On the other hand, insufficient reinforcement fails to develop sufficient strength. The width of a butt weld should be approximately one-eighth to one-fourth inch more than the face of the groove.

Fillet Weld Specifications

A fillet weld is measured by the length of its shortest leg if the weld has a flat face. If the weld has a concave or

convex face, the size of the weld is really designated by the leg length of the largest isoceles right triangle which will fit within the cross sectional contour of the weld.

The convex fillet weld causes an uneven distribution of stresses. On the other hand, the concave fillet weld minimizes the abrupt change of contour and gives better stress distribution, but it involves the deposition of excess metal. For most practical applications, the flat fillet weld or the concave fillet weld is used.

TO DETERMINE THE SIZE OF A CONVEX FILLET WELD

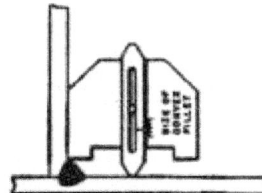

PLACE SAOC AMMST THE TOE OF THE SHORTEST LEO Of THE FILLET " JOE POINTER OUT UNTIL IT KJCHES STRUCTURE AS SHOWN.

READ "SIDE OP CONVEX FILLET"

TO CHECK THE PERMISSIBLE TOLERANCE OF CONVEXITY

AFTER THE SIZE OF A CONVEX WELO HAS BEEN DETERMINED, PLACE ANO SLIDE POINTER UNTIL IT 1 THE FACE OF FILLET WELO AS 1

THE MAXIMUM CONVEXITY SHOULD NOT BE GREATER THAN THAT INDICATED BY "MAXIMUM CONVEXITY" FOR THE ! OF FILLET BEIN8 I

TO DETERMINE THE SIZE OF A CONCAVE FILLET WELO

PLACE BASE A8AINST STRUCTURE AND SLIDE POINTER OUT UNTIL IT

TOUCHES THE FACE OF THE FILLET WELO AS SHOWN.
ON FACE OF°0A0E C °" C * VE F,LL£T "
TO CHECK THE PERMISSIBLE TOLERANCE OF REINFORCEMENT

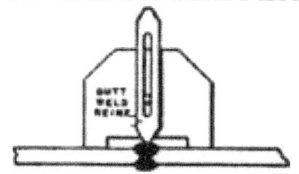

PLACE GAGE SO THAT REINFORCEMENT WILL COME BETWEEN LEOS OF BASE AND SLIDE POINTER OUT UNTIL IT TOUCHES THE FACE OF WELO AS SHOWN.

THE PERMISSIBLE TOLERANCE OF REINFORCEMENT IS THAT INDICATED ON THE FACE OF GAGE

Figure 75.—Instructions for using butt and fillet weld gage.

The correct size of a fillet weld can be determined most easily by the thickness of the thinest sheet or plate being welded. The leg of a fillet weld should be equal in length to V/ 2 times the thickness of the thinest sheet or plate with the following reservation. When there is a wide variation between the thickness of the metal being welded, it is sometimes advisable to use the average thickness of the sheets. Instructions for using a butt and fillet weld gage are explained in figure 75.

TECHNIQUES OF POSITION WELDING

Each time the position of a welding joint or the type of joint is changed, it may be necessary to change any one or combination of the following: current value, electrode, polarity, arc length, and welding technique.

Current values are determined by the electrode size as well as the welding position. Electrode size is governed by the thickness of the metal and the joint preparation, and the electrode type by the welding position. Manufacturers specify the polarity to be used with each electrode. Arc length will be controlled by a combination of the electrode size, welding position, and welding current.

As it will be impractical to cite every possible variation occasioned by different welding conditions, only the information necessary for the commonly used positions and welds is discussed here.

DEPOSITING BEADS

For depositing flat beads the electrode should be held at the starting point for a short period after striking the arc to permit fusion to occur. After fusion has taken place, the electrode should be advanced along the desired weld path at an angle of about 15° with the vertical, thereby depositing a bead of weld metal.

The rate of advance must be timed to allow the work metal to melt and proper fusion to take place. Too rapid an advance will make a thin, irregular bead with questionable

fusion. Too slow a rate of advance may produce overlap , at the edges of the bead.

When depositing horizontal beads in the flat position, no tendency exists for the molten metal to run more to one side than the other. When depositing horizontal beads on a vertical surface, this condition does exist because the force of gravity tends to cause the molten metal to run downward. As a result, special care must be taken to produce a uniform bead without excessive roll or overlap.

In order to produce the proper bead, a short arc should be used and the welding current should be slightly lower than that required for depositing beads in the flat position. The electrode should be pointing toward the deposit at an angle of 15°.

Depositing vertical beads on a vertical surface has the same disadvantage as depositing horizontal beads on a vertical surface—the molten metal tends to run downward. As in making the horizontal weld, a very short arc is necessary, and the proper position of the electrode is essential.

Vertical beads may be deposited upward or downward, although uphill welding is generally preferred. When welding upward, the electrode should be held at approximately right angles to the plate surface. A welding current slightly lower than that for "flat" position welding is recommended.

A wide bead may be deposited by moving the electrode from left to right a distance no greater than four times the diameter of the electrode. A slight upward motion should be used at the end of each stroke of the weave. When making a wide bead, a slightly lower current will be required.

Downhill welding requires the electrode to be pointed upward at an angle of about 15°. A short arc is required and the welding current should be slightly higher than when welding uphill. Weaving may be accomplished when welding downhill if care is taken to prevent the metal from running ahead of the arc. A simple back-and-forth motion is satisfactory. A bare or lightly coated electrode is preferred for downhill welding.

Depositing overhead beads is one of the most difficult processes in the art. The deposited molten metal has a tendency to run or fall downward as a result of the pull of gravity, and it consequently demands great skill to deposit a smooth and uniform bead.

The major portion of metal transfer from an electrode occurs by the formation of drops. Therefore, the welding technique must be such as to transfer the molten drops from the electrode to the plate being welded.

Maintaining a short arc is essential to the deposition of metal in the overhead position. If a long arc is held, difficulty will be encountered in obtaining proper metal transfer and in preventing the deposited metal from running downward in the form of drops. The electrode should be pointed at approximately right angles to the plate and toward the weld at an angle of about 15°.

Most overhead welding is made in stringer beads although a slight weaving motion can sometimes be used. Excessive weaving must be avoided because it tends to create a large pool of metal which is difficult to control in the overhead position.

BEAD WELDING

Welding a square butt joint by means of stringer beads involves the same techniques of welding as depositing stringer beads on a flat metal surface. The techniques for welding beads in different positions were explained in the preceding paragraphs, and they apply to welding a square butt joint.

Square butt joints may be welded in one, two, or three passes. If the joint is welded with the deposition of one stringer bead, complete fusion is obtained by welding from one side. If the thickness of metal is such that complete fusion cannot be obtained by welding from one side, the joint must be welded from both sides.

When the metals to be welded are butted squarely together, two passes are necessary. If the metals must be spaced, three

238500° .-»:< 15

passes are required to complete the weld. In the latter case, the third pass is made directly over the first and completely envelops it.

It must be constantly kept in mind that beads, either of the stringer or weaved type, are used to weld all types of joints. Even though the bead may not be deposited on the same type of surface, its action in the different welding positions and joints is basically the same as its actions on the surface of flat metal. The same fundamental rules apply regarding electrode size and manipulation, current values, polarity, and arc lengths.

The technique of depositing beads in the different types of welding joints and positions is explained, however, to emphasize the additional techniques required to weld them properly.

GROOVE WELDING

Groove welding may be executed in either a butt joint or an outside corner joint. An outside corner joint corresponds to a single V butt joint, and the same welding technique is used for both. It is for this reason that these two types of joints are classified under the heading of grooved welding. There are certain fundamentals which are applicable to groove welds regardless of the position of the joint.

Groove welds are classified as either single groove or double groove. This holds true whether the shape of the groove be a V, U, J, or any other form. Regardless of the position in which a single groove weld is made, it can be welded with or without a backing strip. If a backing strip is used, the joint may be welded from only one side. When a single groove weld is made without a backing strip, the weld may be made from one side, if necessary, although welding from both sides assures better fusion. The first pass of the weld deposit may be from either side of the groove. Complete fusion at the root of the joint may be obtained by chipping the back of the weld to sound metal and rewelding after two or more layers are deposited in the groove.

The number of passes required to complete a weld is determined by the thickness of the metal being welded and the electrode size being used.

Double groove welds are welded from both sides. This type of weld is used primarily on heavy metals to minimize distortion. This is best accomplished by alternately welding from each side—depositing a bead from one side and then from the other. However, this necessitates turning the plates over several times (six times for %-inch plate).

Distortion may be effectively controlled if the plates are turned over twice, as follows: Weld half the passes on the first side, turn the plate over and weld all the passes on the second side, then turn the plates over and complete the passes on the first side.

The root of a double groove weld should be made with a narrow bead, care being taken to insure that the bead is uniformly fused into each root face. When a few passes have been made on one side, the root on the opposite side should be chipped to sound metal to make the groove, and then welded with a single bead weld.

Any groove weld made in more than one pass must have the slag, spatter, and oxide carefully removed from all previous weld deposits before proceeding with the weld deposits over them.

Groove Welding Positions

The first bead deposited in groove welding in the flat position is made in the same manner as running a bead on a flat metal surface. If additional passes are necessary to complete a groove weld, they may be deposited in the form of stringer beads or weaved beads. The finish weld pass is usually weaved, although such technique is not mandatory.

Figure 76 illustrates groove welding in the flat position of single and double groove joints. These processes are included to demonstrate possible bead sequence of making deposits. Small diameter electrodes are preferred, especially for the first passes in a multiple pass groove weld.

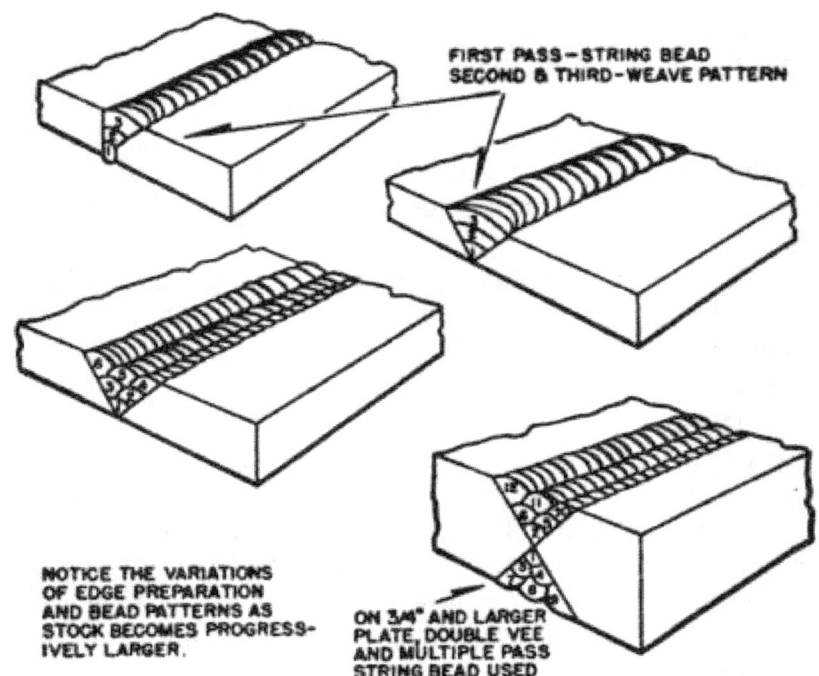

Figure 76.—Greov* welding (flat petition) bead sequence.

A horizontal groove weld, either between vertical plates for a butt joint or in a horizontal corner joint, is one of the most difficult welds to make. A number of joint designs are used, but in all cases they should be proportioned so that the kerf angle of the top plate is not too small.

The bead sequence should be as shown in (A) of figure 77. Notice that beads 3 and 5 are practically deposited in the overhead position, and must be welded as such. This is an additional reason for putting a wide bevel on the top plate—no less than 45°—because it permits better fusion to the top plate.

The welding joint may be such that a single bevel would be inadvisable, in which case the horizontal groove weld may be made as illustrated in (B) of figure 77. The groove welds shown in this drawing are of the single groove type.

If a double groove weld is to be made, the welding technique described would be applied to both sides.

Regardless of the welding joint shape, a horizontal groove weld should be made by using a series of stringer beads or weave beads. Small diameter electrodes are preferred, especially for the first passes. A short arc must be maintained at all times and care taken to prevent roll and overlap. Depositing metal on the top kerf surface approximates overhead welding, and consequently special care must be taken to obtain complete fusion.

The number of passes required to complete a vertical groove weld will depend upon the thickness of the metal. The direction in which the passes are made is dependent on a number of factors.

If welds are made by welding downhill, more passes are required, in a multiple pass weld, and the result is greater distortion, slightly higher tensile strength, lower ductility, and more porosity. For this reason, downhill welding is

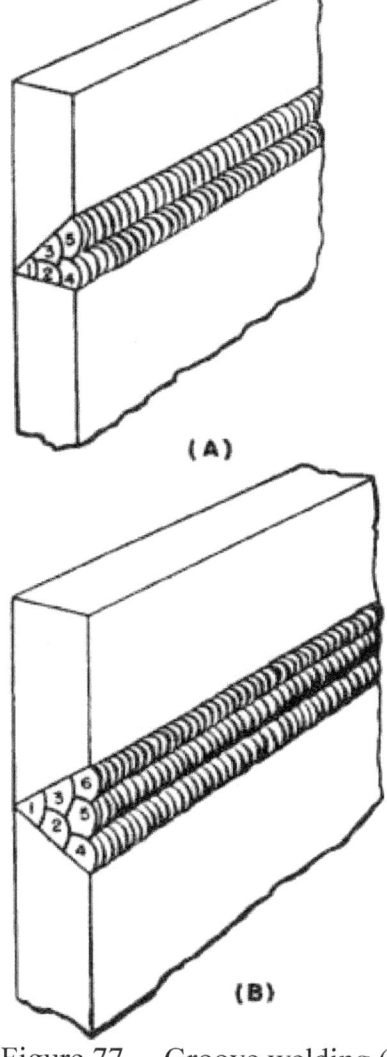

Figure 77.—Groove welding (horizontal position) bead sequence.

seldom used except on very thin gage metal, or on the first pass of a heavy weld. Downhill welding on the above is performed to eliminate the danger of burning holes through the plates.

Multiple pass vertical groove welds may be welded with the first pass made downhill in order to seal the joint and prevent burning through the edges at the joint. The deposition of the second pass will almost completely remelt and fuse the first pass, thereby insuring a sound joint the full thickness of the plate.

As previously mentioned, downhill welding requires more welding current for a given size electrode than does uphill welding. As a result, if a smaller diameter electrode is used on the downward pass than in the upward pass, the same current setting will be found satisfactory. This eliminates the necessity for changing the current setting between passes.

Vertical groove welds are usually best made by using a weaved bead, with the exception of the first pass in a multiple pass weld, which is a stringer bead. When a vertical groove weld can be completed in one pass, it is recommended that the weave be executed in an approved manner. A number of weaving procedures can be used, and the type preferred is dependent somewhat upon the welder's skill. One procedure consists of a simple triangular weave. In cases where trouble is experienced from undercutting or excessive local heating at the weld, a whip procedure may be used. On the upward motion of the electrode, the arc is drawn out, thereby allowing the deposit to cool slightly. The upward whipping motion of the electrode should extend about 1 to 2 inches above the deposit.

The weaving procedure recommended on the upward passes of a multiple pass weld consists of passing the electrode across the face of the weld and applying a slight upward and outward motion at the plate surfaces. This is done in order to obtain complete fusion and to push the molten metal into any undercut portion that might have been obtained.

Proper control of the arc is essential to insure complete

penetration and fusion and to prevent undercutting at the edges of the weld. A uniform weave and rate of advance is important in obtaining a weld of uniform quality and appearance. A short arc is required at all times.

A groove weld in the overhead position is best made by depositing stringer beads in the groove joint. In some cases a weave bead may be used, but this is not advised for the welding beginner.

The most difficult task in making an overhead groove weld is that of obtaining complete fusion in the first layer without burning holes in the thin edges of the plates at the root of the weld. Periodical lengthening of the arc or whipping the electrode along a joint when the metal appears to become too hot will greatly aid in preventing the burning of holes.

A short arc should be maintained to facilitate metal transfer.

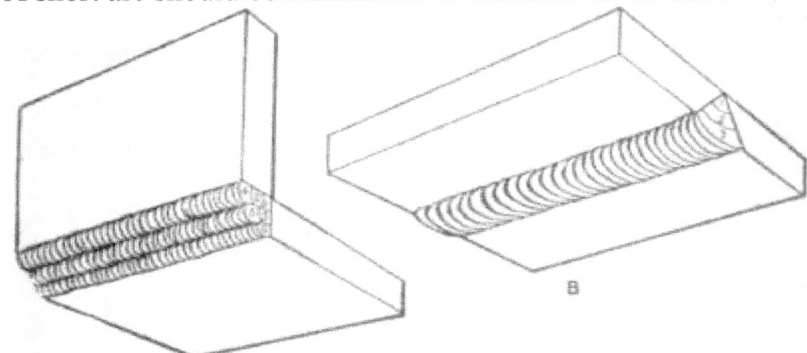

A

Figure 78.—Groove welding (overhead position) bead sequence.

Figure 78 illustrates two groove welds with bead sequence indicated. (A) in figure 78 shows stringer bead technique, while (B) demonstrates weaved bead technique.

FILLET WELDS

A fillet weld may be made in one or more passes. The number of passes is governed by the electrode size and metal

thickness. The size of a fillet weld that should be made in a single pass will depend upon the diameter of the electrode, the thickness of the plate being welded, and the rate of electrode advance.

The maximum size of single pass fillet welds that should be made with a given size electrode is as follows:

%-inch electrode %-lnch fillet weld.

%2-inch electrode %-lnch fillet weld.

%8-inch electrode %«-inch fillet weld.

When making a fillet weld in a lap joint, or its equivalent, the electrode should be held at an angle bisecting the angle formed by the metal being welded. It should also be pointed backward.

When a fillet weld is being made in a tee joint, it is desirable to point the electrode backward toward the weld and at an angle of 30 degrees with the horizontal.

i

Fillet Weld Petitions

To complete a fillet weld in the flat position in one pass, a stringer bead may be deposited, or the electrode may be weaved slightly.

This weaving motion builds up the metal on the vertical surface and widens the area of fusion on the horizontal surface. Care must be taken to obtain complete fusion at the root of the weld. Too high a welding current may cause the vertical surface to melt away, thus resulting in an undersize fillet.

In a two or more pass fillet weld, the first bead is deposited as a stringer bead by moving straight down the joint without weaving. Care must be taken to obtain complete fusion at the root. The weld may be completed by depositing additional stringer beads or a weaved bead or beads. Figure 79 illustrates the sequence for depositing beads in a lap or T-joint. This figure is purely illustrative and does not signify that all lap joints or T-joints must be made in three passes. Many additional passes will be required for some

Hgura 79.—S«qv«K« for depositing b*ads in a lap or T-jolnt.

fillet welds. In each case, however, the beads are deposited on the horizontal surface first. (4) in this illustration demonstrates bead sequence in a lap joint, and (B) shows the same sequence in a T-joint.

Fillet welds made with stringer result in a higher quality weld with less tendency to undercut.

Flat fillet welds and horizontal fillet welds are to all practical purposes the same type of weld and require identical techniques.

The production of a fillet weld in the vertical position is very similar to the production of a vertical groove weld— such as a single V butt weld—which has been previously explained. Consequently, the same weaving procedures will apply. The major differences between the two types of welds lie in the fact that the root of the weld is deposited on heavy sections rather than on thin beveled edges. This makes complete penetration at the root without burning through the plates somewhat easier to obtain.

The vertical fillet weld may be completed in one or more passes. In a 'multiple pass weld, downward welding is usually preferred for the first pass because it seals the crack between the plates and eliminates a great deal of trouble from arc blow. The deposition of the remaining passes may be made by using a triangular weaving technique, or a series of beads may be used. If the weaving technique is used, the thickness of the weld and the final size of the fillet can be determined to a certain extent by the rate you advance the weave.

A fairly short arc should be held, and special care taken to prevent undercutting at the weld edges. Figure 80 illustrates fillet welds in a vertical position.

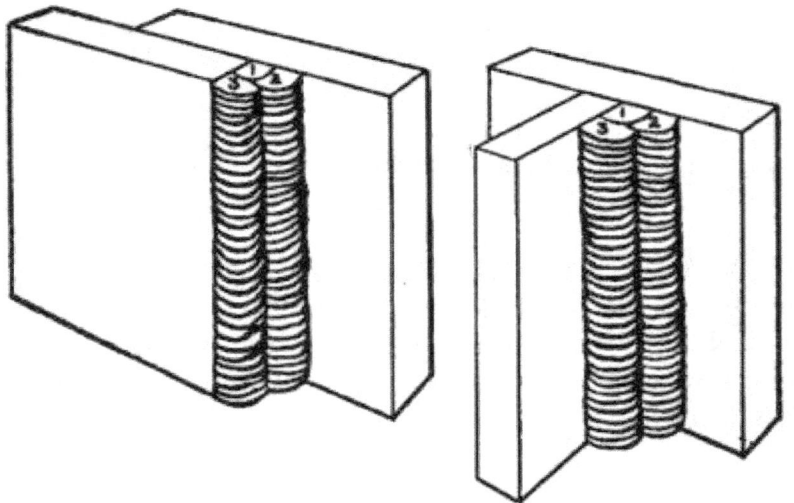

Figure 80.—Rile! welding (vertical position).

Overhead fillet welds may be made in two positions.. A fillet weld made with the plates positioned at an angle, as shown in this illustration, is similar to an overhead groove weld and should be made as such.

Overhead fillet welds are generally made in a series of string beads similar to the method discussed for overhead groove welds. Because of the heavy plate sections at the root of the weld, however, there is no danger of burning holes in the parts. A short arc should be held at all times.

QUIZ

1. What is freezing?
2. What are two methods of striking the arc ?
3. What must be done to form a uniform bead ?
4. Where should the arc be restarted?
5. What is weaving and why is it used?
6. What is multiple pass welding?
7. What governs the number of passes in a fillet weld ?
8. When the position of a joint or the type of the joint is changed, what may have to be changed in the welding process I

CHAPTER 11
INERT ARC AND ATOMIC HYDROGEN WELDING THE INERT ARC WELDING PROCESS

Inert arc welding is a method for fusion joining of metal by an electric arc, in which the arc is struck and maintained in a shield of inert, or inactive gas.

The more familiar arc welding requires a flux to prevent oxidation of the weld during the process, as the molten flux dissolves oxides and nitrides and floats away impurities. This flux on heavily coated electrodes also creates a gas shield, or envelope, which reduces oxidation of the hot metal due to contact with normal atmosphere during the welding process. By using an inert gas, it becomes unnecessary, in most

cases, to use a flux, thereby eliminating cleaning of the finished weld.

INERT GASES

An inert gas is one which is chemically inactive.

Although there are several inert gases which may be used for welding, helium and argon are the two most common. Helium gas can be used as a shield for most jobs, but is very light and does not surround the work as effectively as does argon gas, which is heavier. Argon gas consumption is about one-third less than helium because of its greater density. Being heavier than helium, it has better enveloping qualities.

HELIUM AND ARGON TESTS

Inert gas welding tests were made at the Naval Aircraft Factory as this text was being written, on a variety of alloys, using both helium and argon gases. Welds were performed with a Linde Heliarc Torch No. HW4 and a Rexarc alternating current welding machine manufactured by the Siglet Feed Generator Co. The results of these tests are shown in table 19.

The helium tests in the table were made with 99.8 percent pure helium shipped specifically for this purpose from the Amarillo Helium Plant at Amarillo, Texas. The argon gas used was the standard commercial argon supplied by the Linde Air Products Co. for inert gas welding.

These tests demonstrated that the aluminum alloys are readily welded with argon gas but do not weld readily with the helium gas unless a flux is employed. In welding magnesium, a better weld was produced with helium than with argon on thin gage material, but heavier gage material welds as well with argon as it does with helium. Phosphor bronze welded satisfactorily under argon but could not be welded under helium.

On 18-8 steel sheet, argon and helium appeared to weld with equal facility. However, in welding a 30° angle butt with 0.028 wall tubing, a satisfactory weld could not be

obtained with helium, whereas the tube welded readily with argon. In welding Inconel sheet, the heavy gages welded readily with either gas, but thin gages could not be welded satisfactorily with helium, while argon produced good welds.

During the tests, a tank of double charcoal purified helium was received from the Amarillo Helium Plant. Double charcoal purified helium (in accordance with BuAer Letter Aer-SU-2 F20 84635 of 18 November 1948) is designated as grade A helium. It is practically 100 percent pure. Welds were made with this grade A helium on aluminum, phosphor bronze, and thin-gage Inconel, which previously had been difficult to weld with the 99.8 percent (grade B) helium. The grade A helium made these welds satisfactorily, and is considered far superior to the grade B welding helium.

In a specific job which involved the welding of 3S aluminum parts that did not fit perfectly, it was very difficult to obtain a good weld with grade A helium, whereas the job was easily welded with argon. However, when a flux was used, a good weld was obtained with helium. As an alternative, argon was mixed with helium in the proportion of one part argon to three parts helium, and the weld was accomplished with no difficulty.

These tests disclose that while satisfactory welds can be made on thin gage metals (less than 0.020) with grade A helium, they are accomplished with much more difficulty than when argon gas is used. One of the objectionable difficulties is the strong tendency to burn through thin metals because of the greater heat developed when using helium than when using argon. Another important factor is that a much more stable arc is produced under argon than helium, which results in a better quality weld with argon. The effects of these factors are also evidenced in the greater irregularity of the underbead side of a helium weld as compared to that of an argon weld.

In adjusting welding conditions to produce the best quality weld under each of these gases, it was found that welding

under helium is much more sensitive to electrode size than welding under argon. Argon has the further advantage of being about 10 times as heavy as helium which, among other things, is believed to help in maintaining more intimate contact with the work.

Tab!. 19.—QUALITY OF INERT GAS WELDS

G—Good; VO-Very good; NO—No good; F-Fair; P—Poor. •Could be welded with flux. ••30° butt weld.

The tests led to the conclusion that double charcoal purified helium (grade A) is an excellent medium for inert gas welding and, in general, can be broadly applied.

INERT ARC PRINCIPLE

The principle of the inert arc is the same as the conventional electric arc discussed in the preceding chapters. The arc is actually the passage of electricity through atmospheric resistance between the electrode and the bare metal. An arc in argon gas has less resistance heat than in helium, and is similar in appearance to an acetylene welding flame. The arc in helium is wider spread and more ball-like in shape.

This process is termed "arc in inert atmosphere" by AN Specification, "Tests, Aircraft Welding Operators Certification", and is also variously known as Heliwelding, Heliarc Welding, Argon Gas Welding, and Inert-Gas-Shielded Welding.

EQUIPMENT

Figure 81 illustrates standard inert arc welding equipment. Two important points to keep in mind while studying information presented in this chapter are that tables referred to are from specific manufacturers and should not be assumed to be all-inclusive in content. The functions of similar component parts of different makes of machines are identical although they outwardly may not appear to be so.

Power Unit

The power unit supplies alternating current through a transformer, or direct current through a motor generator.

A suitable source of electrical power—either 220 or 440 volts—and a water supply if a water-cooled torch is used, are necessary.

The ground lead from the welding machine should be

238500° -53— HI

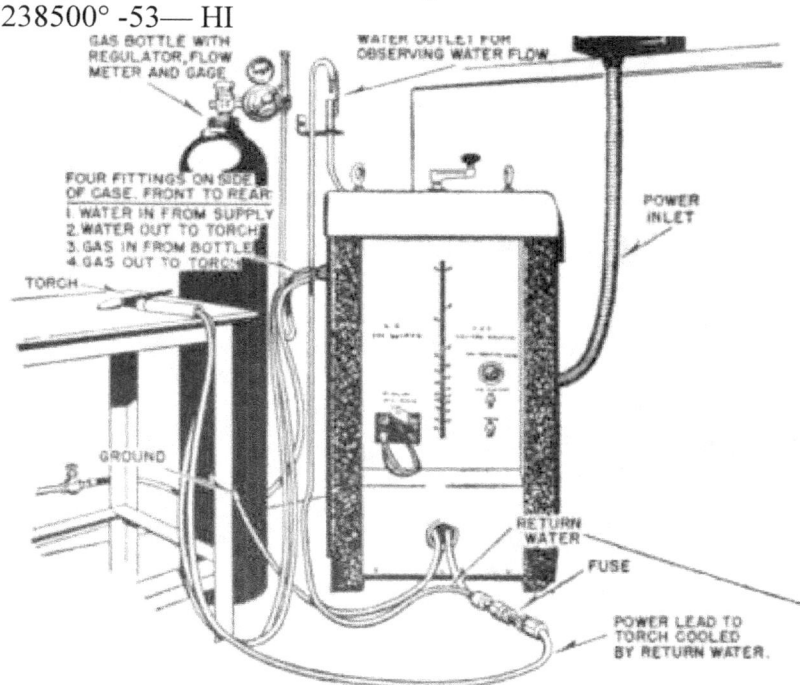

Figure 81.—Typical inert arc welding equipment.

clamped to the work table, which in turn should be connected to an earth ground.

There are two high frequency adjustment points above the high frequency unit. This spark gap installation is used only in alternating current machines for high-frequency stabilization to aid in striking and maintaining the arc. In other words, it forms a "bridge" to smooth out or provide a more even flow of current.

Torches

Output will range from 10 to 500 amps a. c. and 800 amps d. c. in hand torches, and 500 to 1,000 amps in machine welders, depending upon the work to be performed, as shown in tables

20 and 21.

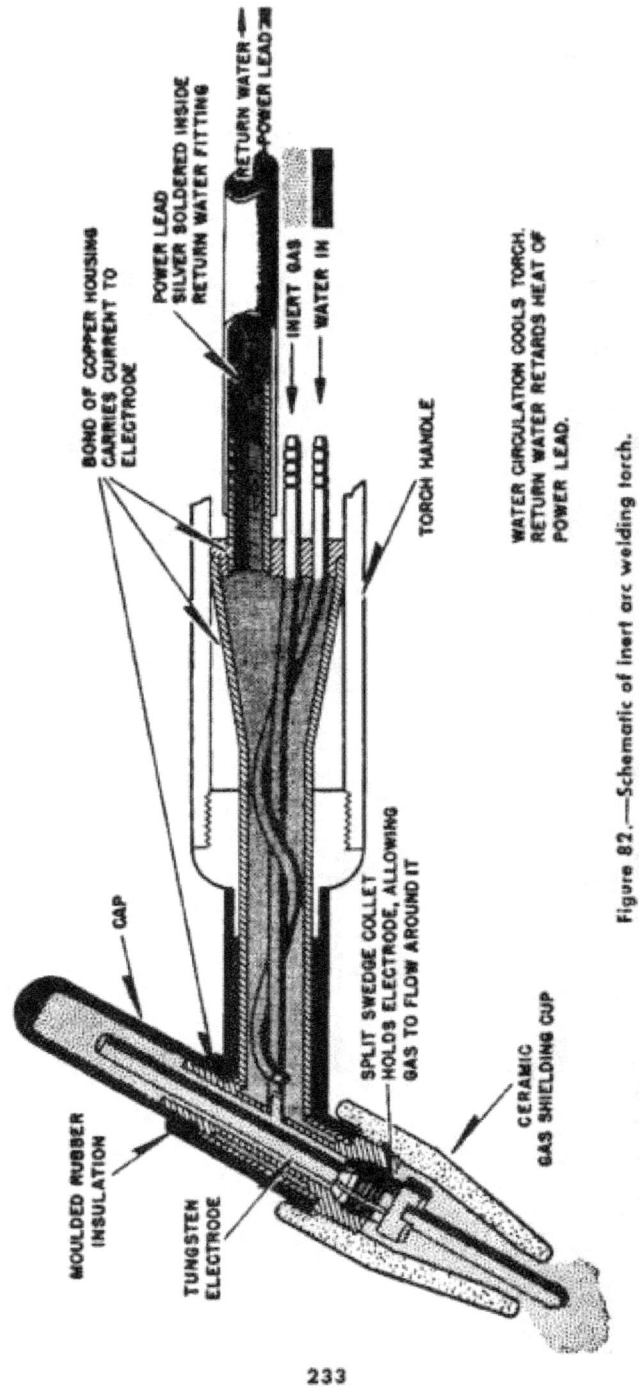

RETURN WATER
POWER LEAD

POWER LEAD
SILVER SOLDERED INSIDE
RETURN WATER FITTING

BOND OF COPPER HOUSING
CARRIES CURRENT TO
ELECTRODE

INERT GAS

WATER IN

TORCH HANDLE

WATER CIRCULATION COOLS TORCH.
RETURN WATER RETARDS HEAT OF
POWER LEAD.

CAP

SPLIT SWEDGE COLLET
HOLDS ELECTRODE, ALLOWING
GAS TO FLOW AROUND IT

CERAMIC
GAS SHIELDING CUP

MOULDED RUBBER
INSULATION

TUNGSTEN
ELECTRODE

Figure 82.—Schematic of inert arc welding torch.

233

The water-cooled torch (fig. 82) is designed for gas-shielded arc welding using either alternating or direct current with straight or reversed polarity.

Water requirements for water-cooled torches are 1 to 2 pints per minute for hand welders, and 2V£ to 4 pints per minute for machine welders. Sufficient pressure should be available to provide cooling circulation.

Water cooling of torch and cable makes possible the use of a light-weight power cable capable of carrying the full output current without danger of overheating. A special fuse of about 45 amps is installed in the power lead line to the torch to protect equipment in case of water stoppage.

The torch head, handle, and gas cup are well insulated to protect the operator from shock, and to prevent damage to the work by accidental arcing.

Inert gas, cooling water, and electrical power lines all lead into the torch through the insulated handle. The tungsten electrode is encompassed, in the torch head, by a tube through which the gas passes. This gas-conducting tube is in turn surrounded by a water jacket.

The electrode is snugly held in position by a four-fingered replaceable clamp, which comes in various sizes to hold electrodes from y 16 - to %-inch in diameter.

Gas-shielding cups are made of ceramic tile and selected according to the size of the electrode to be used. These units should be secured finger-tight only.

Table 20.—ELECTRODE AND CUP SIZES

Courtesy Linde Air Product! Co.

Electrodes

Electrodes are commercially pure tungsten (99.9 percent), and have a very high melting point They are practically nonconsimiable—normal consumption is 0.001 to 0.005 inch per hour—although accidental contact with the work will cause them to burn.

If the electrode comes in contact with the base metal or the work table, a small bead or ball will appear at the contact end, which causes the arc to become unstable. This ball is removed by grooving the. electrode just above the fouled point with a file, then breaking it off with pliers. It is essential that the electrode be grooved to prevent splitting when the burned metal is broken off.

Loss of the electrode due to corrosion can be somewhat prevented by leaving the gas on a short while after the arc has been broken to allow shielded cooling. Some machines are designed to automatically provide delayed gas shut-off.

Electrodes are available in diameters of one-sixteenth to three-eighths inch and in length from 3 to 12 inches. The

Table 21 .—CURRENT RANGES FOR ELECTRODE SIZES

Welding current (amperes)

Electrode diameter

(inches)

Gas cup (number)

Alternatini
lternating cur-
rent with high
frequency

Direct current straight polarity

Direct current reverse polarity

He

X-

X-

8 (H")~

6 <%")--6, 7,8...

6, 7, 8...

7, 8

100-180. 150-240*..

250*
Do not use
40-120
75-150
100-250*
250*
Do not use
Do not use
10-20 15-30 25-40 40-80 80-125
•Use water-cooled gas cap.

Courtesy Lmde Air Products Company

diameter of the electrode to be used is determined by the current which is governed by the type and thickness of the material to be welded. In general, the electrode diameter should be increased one-sixteenth inch for each 100 amps to be used. Table 21 furnishes more exact current ranges for electrode sizes, but should be referred to only as a basis for more accurate adjustments as determined by experience.

Inert Gas Cylinders

Argon gas cylinders are of 242-cubic-foot capacity, and are charged to 2,200 p. s. i. When withdrawal of gas lowers the pressure to 25 p. s. i., the tank is considered empty and must be processed before refilling.

Argon cylinders contain 99.9 percent argon and 0.01 percent nitrogen. Navy-type argon cylinders are painted gray with one or two white bands at the top. The type with one white band is water pumped and is the kind used in inert-gas shielded-arc welding. Helium cylinders are painted gray with either one or two orange stripes at the top. The type of helium in the cylinders with two orange stripes is the oil-free type used in welding. The same care should be exercised in handling and storing argon and helium bottles as practiced with acetylene and oxygen containers.

Miscellaneous Equipment

The regulator (fig. 83) employed for inert gas welding is identical in design and construction to the conventional oxygen regulator, and performs the same function of reducing high cylinder pressure to that required at the torch.

Flowmeters attached to the regulator indicate gas flow to the torch in liters per minute or cubic feet per minute.

The flowmeter should always be attached in the cylinder gas line after a regulator is installed. Crack the cylinder valve before installing regulator, flowmeter, and gage in order to blow out all foreign matter.

It is essential that cylindrical type flowmeters be mounted exactly vertical. Stand clear of the valve opening when cracking the cylinder valve.

The STANDAKD ARC WELDING HELMET, with NoS. 10 to 13

lense, is absolutely necessary. The use of gloves and protective clothing to prevent exposure of all body surfaces to the arc while welding, is equally important.

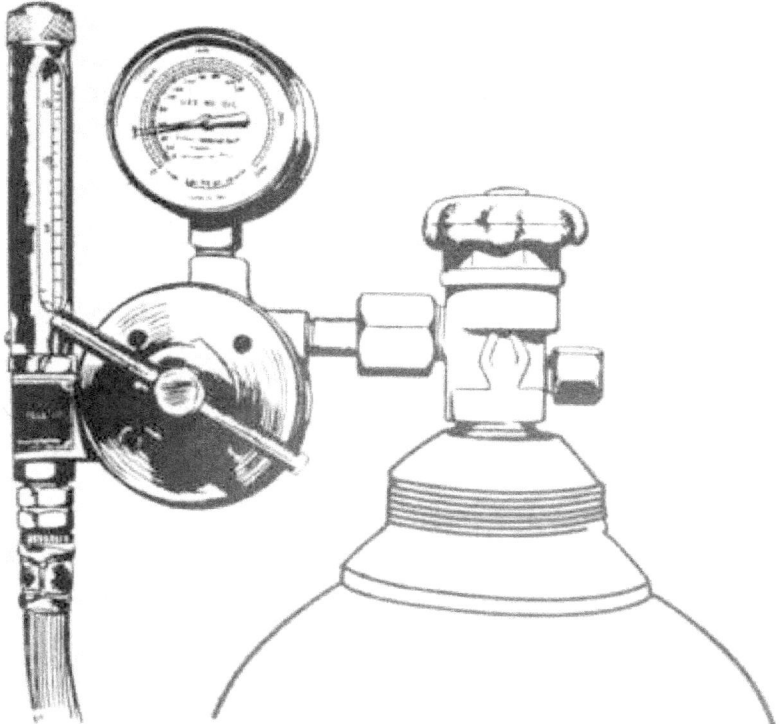

Figure 83.—Regulator, flowmeter, and gage installation.

Welding rod may be any high quality type as recommended for other types of welding a particular type of material. Reinforcement or weld metal additions should, in most cases, be the same as the base metal.

Stainless steel filler rod should have about three percent higher carbon content than the parent metal to replace that portion burned out during the welding process.

Flux is not generally required when welding with inert arc, although there are a few exceptions. When welding aluminum with direct current, straight polarity, it is necessary to use aluminum flux to obtain fusion.

Optional Equipment

Foot-operated rheostats are available for installation in the field circuit to change the arc for varying thickness of material. It provides a more convenient method for making closer current settings while welding, and also allows the operator to strike, maintain, and break the arc more smoothly.

A water-oas BHTJTOFF valve controls the flow of water and gas to the torch. Both water and gas are shut off when the torch is hung on the hooked-arm provided for that purpose.

WATER LEAKS AND WATER STOPPAGE

In case of water leakage, discontinue welding. Two possible causes of leakage in water lines are excessively high water pressure and improperly sealed hose connections.

When hose is damaged near the end, it is necessary only to cut away the broken section and reinstall it to the fitting. Rubber cement should be applied to the nipple and allowed to become tacky before securing to the hose. Attach the hose to the connection with about three turns of No. 16 or No. 18 copper wire.

Water stoppage may be caused by an accumulation of dirt in the small passages of the hose fittings in the torch. This condition can usually be corrected by disconnecting the water lines and reconnecting to reverse water flow.

SETTING UP FOR THE JOB

Before setting up for a job, make a check on the equipment needed. In general, the items listed below will fill the bill for most jobs.

A. c or d. c. welding machine. Inert arc torch. Source of inert gas.

Source of running water (if required). Hose (suitable color or transparent). Regulator (single or double stage). Gas flow meter (s).

Water flow meter or visible flow of water. Tungsten electrodes of suitable size (polished). Gas cups for corresponding sizes of electrodes. Protective clothing. Leather gloves.

Head shield with No. 10-13 lens.

Rubber floor mat.

Pliers and tongs.

Clamps, backing plates, or jig.

Flux, if required.

Extra fuses (45 amperes).

Filler rods, if required.

After material and equipment are assembled and set up in a welding booth or curtained area, the next step is to select the right type of current to be used as shown in table 22. Tables 23, 24, and 25 list general operating procedures for various types of metals.

If, for example, the job to be done is on 2S, 0.031 aluminum sheet, table 23 shows that high-frequency stabilized alternating current is recommended as first preference because of its excellent operation.

The next thing to determine is current range required for the welding speed. Amperage for 0.031 aluminum sheet is shown to be 20-75. For a given amperage, a specified size of electrode should be used. Remember that an electrode of about Vie-inch diameter for each 100 amps, is generally standard.

It should be recalled that figures given in all these tables are to be used only as a basis from which to work. The actual settings for best results can be obtained only through experience.

Table 22.—SELECTING THE RIGHT TYPE OF CURRENT

Material

High frequency stabilized alternating current

Direct current

Straight polarity

Reverse polarity

Magnesium up to %t inch

Magnesium above #e inch

Magnesium castings

Aluminum

Aluminum castings

Stainless steel 0.015 to 0.050 inch. Stainless steel 0.050 inch and up._ Low-carbon steel 0.015 to 0.050

inch (killed steel only)

Low-carbon steel 0.050 inch and

up (killed steel only)

High-carbon steel 0.015 to 0.050

inch

High-carbon steel 0.050 inch

and up
Cast iron
Deoxidized copper to 0.090 inch-Brass alloys
Everdur__
Monel* _..
Silver
Hastelloy alloys
1 1 1 1 1
1 2
2 2 2
1 2 2 2 1
N. R. N. R. N. R. N. R. N. R.
2 1
2
2
N.R,
1 1 1
1 1 2 1 2
1— Excellent operation—first preference.
2— Good operation—second preference. N. R.—Not recommended.
•Heliarc process not always satisfactory for
Table 23.—GENERAL OPERATING DATA FOR ALUMINUM
Aluminum
Ctmrtety Linde Air Products Co.
Tablt 24.—GENERAL OPERATING DATA FOR MAGNESIUM
OmrUtf Linde Air Product* Co.
Table 25.—GENERAL OPERATING DATA FOR STAINLESS STEEL
•2 passes.
Courtesy Lmde Air Product* Co.

Regardless of the metal being welded, joints and adjacent surfaces must be thoroughly cleaned. Filler rod if used should also be cleaned. Oil or grease should be mechanically or chemically removed. An abrasive disc sander is recommended for mechanical cleaning. Chemical cleaning should be performed using a caustic bath followed by a sulphuric acid brightening bath. Do not use a wire brush or nitric acid, as this will contaminate the weld.

STRIKING THE ARC

Before striking the arc, see that the rubber mat is in place, so that the deck and feet may be kept dry.

To weld aluminum without flux, a. c. power must be used. If d. c. straight polarity is used, flux must be applied to the underside of the work to obtain fusion.

The electrode is about y 16 -inch diameter for every 100 amps, and about 80 amps is required for each y 16 -inch thickness of material to be welded.

If high frequency stablizing current is used, it is possible to start the arc without touching the work. Set the argon

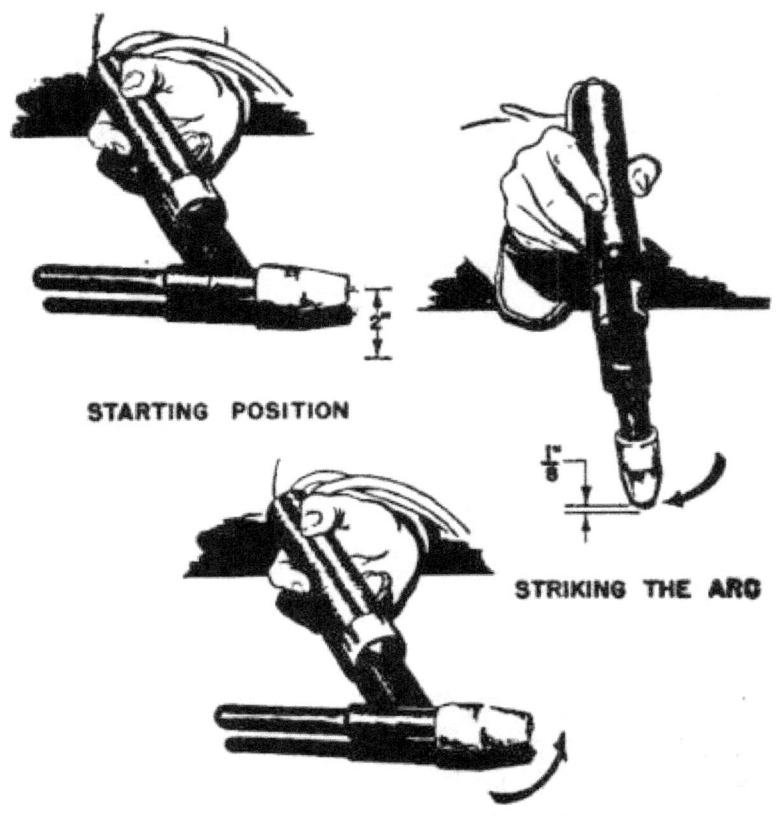

STARTING POSITION

STRIKING THE ARC

BREAKING THE ARC

Figure 84.—Striking and breaking th« arc 244

flow to read about 4.4 liters per minute (9.35 cubic feet per hour), and the current range to 20-75 amps.

The torch handle is held in any comfortable manner and steadied with the other hand, if necessary. Hold the torch head in a horizontal position about 2 inches above a piece of heavy scrap steel or carbon. Allow the gas to flow for a few moments to purge the line of any impurities. Pull down the protective helmet and proceed as follows, referring to figure 84.

Swing the cup down so that the end of the electrode is approximately one-eighth inch from the surface of the block of scrap metal and the torch is in a vertical position, as shown in figure 84. Notice that this is a backward swing of the electrode. Practice this motion until the electrode can be brought down to a starting position, and the arc struck and maintained without touching the work. Observe that as the electrode becomes hot, the arc is more easily established at a greater distance from the work. As the operator gains speed in striking the arc, he will find it helpful to warm up the electrode on scrap metal before proceeding with a job.

One good reason for learning to strike the arc as closely as possible to work without touching it is to give satisfactory gas coverage to the weld surfaces.

Never allow a cold electrode to touch the work surfaces. As the electrode heats it will become necessary to make a faster swing to strike the arc.

If d. c. current is used, it may become necessary to touch the work with the electrode to strike an arc. Use the procedure previously discussed, quickly withdrawing the electrode to about one-sixteenth inch from the work. Touching the work with the electrode to strike an arc should be avoided if possible.

A hot electrode arc is longer than one which is cold, and the longer the arc the more likely it is to mark weld surfaces. This is important because when you stop the arc it must be extinguished quickly.

Stopping the arc is accomplished by swinging the electrode up and away from the work in a "snap" action, as shown in figure 84. This should be practiced along with starting the arc to prevent undue burning of the work.

RUNNING A BEAD

After mastering striking, maintaining, and breaking the arc in a smooth manner, try running a bead without filler rod.

Support a practice sheet of 0.031-2S aluminum with backing bars on the table. Strike the arc about an inch in from the edge of the trial sheet after first warming-up the electrode.

To form a puddle, use a slight circular motion of the electrode about one-eighth inch above the work until the puddle is about one-fourth inch in diameter. Then advance the puddle with the torch held at an angle of about 75°, in much the same manner as in electric arc welding. Practice running beads, both forehand and backhand, until a good, smooth bead pattern is obtained. When breaking the arc, remember to allow gas to flow while the electrode is cooling.

Good penetration is indicated by a very small smooth bead on the reverse side of the work.

Check the bead for black discoloration which may be caused by collection of impurities on the electrode due to accidental contact with the work. If the electrode tip appears to be distorted, pull it out with pliers, groove just above the damaged area, and break off. Readjust to Va-inch extension beyond gas shielding cup. If the electrode is not beaded, but impurities are still suspected, draw several arcs on a heavy scrap piece of steel or a block of carbon. This procedure should remove minor impurities which may become stuck to the electrode.

If the electrode is not a silvery white, or if it has a bluish color, there may be insufficient gas pressure from the tank, a leak in the gas lines, or the gas shielding cup may be loose. It is even possible that the lines were not thoroughly purged before striking the arc or that the gas was turned off before the electrode was sufficiently cooled.

Welding speeds up to 40 inches per minute are possible with inert arc methods due to an almost pin-point concentration of heat in the weld area. There is not as much heat absorbed by the entire piece of work, and correspondingly little warping when the metal cools.

It is possible to weld aluminum of one thirty-second to three-fourths inch by hand without preheat. Pieces %• to 114-inch thickness require a preheat of 300° to 400° F. Thicknesses of 0.005 inch can be very satisfactorily welded by machine.

Preheating can be accomplished by using the torch itself, special preheat ovens, or gas flames directed on the work in such manner that these auxiliary flames do not disturb the inert arc flame. Air currents which disturb the flame or the gas shield are not conducive to good welding results.

USE OF FILLER RODS

If a yi 6 -inch filler rod is not available, shear a few strips from the sheet from which the practice piece was cut.

Strike the arc in same manner as previously described and develop a puddle about an inch in from the edge of the practice sheet. Move the arc to the rear of the puddle and, holding the filler rod at an angle of 15° to the horizontal and in line with the direction of travel, quickly touch the forward edge of the puddle. The filler rod should touch only the forward edge of the puddle, and should not be introduced into the arc stream as this will cause splatter and excessive melting of the rod. Filler metal should be added by short, quick dips of the rod, allowing only a small amount of rod to be deposited as the torch is advanced.

MAGNESIUM WELDING

When using filler metal it may be noticed that the rod kas to be held at a very flat angle and touching the plate, advancing it to the puddle as it is required. Do not use a "dipping" motion as was suggested for welding aluminum. The torch

238500* 53 -17

will appear to u bite v off small pieces of filler metal as it is advanced. It may be found that a very short arc is required and that the torch angle has to be almost vertical, being tipped only enough for you to see the work.

In cases where it is impracticable to apply flux it will be necessary to provide a separate means for supplying gas flow under the weld area. Another gas cylinder can be used or a line can be "teed'* off from the gas hose to the torch. If gas is drawn off from the torch line the gas setting will have to be set proportionally greater, as determined by experience.

Examine the weld for penetration and defects. Improper penetration is indicated by a fine line on the reverse side of the weld. This condition can be prevented by slightly beveling the underside of pieces to be welded. The bevel will let molten weld metal float away oxides formed by heat as it flows down between the sheets.

If cracks appear at the ends of the weld, try preventing recurrence by butting pieces of the same material as that being welded against each end of the joint, clamping them in place. Start and stop the weld on these scrap pieces. It is also possible, in some cases, to start welding in the center of the joint, working out to each end.

When jigs or clamps are employed to hold pieces for joining, keep in mind the comparatively great expansion of magnesium. Stresses of several thousand pounds are sometimes developed in a weld area. Such stresses can be relieved by careful heating.

When performing inside corner welds it may be necessary to increase the electrode extension; but keep the extension to the minimum to insure maximum gas shielding.

STAINLESS STEEL WELDING

Procure a supply of ^-inch, 302 stainless steel practice sheets well squared and properly cleaned. Always use the type of filler rod recommended for the kind of metal to be welded, although strips cut from the parent metal may be

used for certain jobs. For the job we shall discuss, %-inch, Oxweld, No. 28 rod will be used.

Ordinarily, rods containing about 3 to 5 percent more carbon than the weld metal are desirable since the intense heat will burn a little carbon out of the base metal. The amount of chromium and nickel lost will be comparatively small due to the concentration of heat in such a restricted area when welding with the inert arc.

Table 22 shows that %-inch stainless steel gives alternating current high-frequency as second preference. To gain experience in using direct current, straight polarity, and because it is given first preference, let's set up for this preference.

For direct current welding of stainless steel by this method, fit the torch with a % 2 -inch electrode and a No. 6 gas cup, with the electrode projecting about one-eighth inch and hooked up for straight polarity (torch negative, work positive). It is important to select the correct size electrode, as it is difficult to start and maintain the arc with an oversized electrode. Furthermore, an electrode which is larger than that recommended for the current setting is apt to cause an erratic arc which in turn results in difficult puddle control.

Adjust the current setting to 125 amperes and high voltage. Notice that this is lower than the figure shown in table 25, but it is better for beginners to start with lower current settings. Argon flow should be four liters per minute at 15 p. s. i.

If direct current is not available for this practice job, proceed the same as for aluminum and magnesium, referring to the tables for correct settings. The arc is struck in the same manner and generally similar techniques are practiced.

If direct current is used, the work must be touched by the electrode using the same swinging motion as described for striking an arc on aluminum and magnesium. Be sure to withdraw the electrode quickly and thereafter do not touch it to the work while welding.

It will be found that the electrode should be held about one-eighth inch from the work and the torch angle at 75° or more.

Good penetration is shown either by the oxide deposit (color bands) on the underside of the work or a very small, smooth bead. A finished job should have a regular bead pattern, be shiny in appearance, and require no cleaning.

When using a rod, dip it rapidly in and out of the forward edge of the puddle to assure addition of filler metal at a uniform rate.

Oxidation on the underside of work can be somewhat prevented by the use of back-up bars, gas shielding, or by use of stainless steel flux mixed with shellac and painted on the work.

Cleaning is best accomplished by dipping the work in de-* greasing solution and/or pickling for removal of heavy scale.

POINTERS ON JOINTS

When it is necessary to preheat sheet or plate for butt joints, the torch head should be held almost vertical.

Filler rod is not usually required on metals up to *4 6 -inch thickness. Do not add filler metal until puddle is well developed. Welds on metal thicknesses up to one-eighth inch usually require only one pass.

It is best to weld a vertical butt joint from the top down.

Lap joints, when made on aluminum, require a short electrode extension with the torch held at about 45° to the work. The rod must be alternately dipped into the forward edge of the puddle, quickly removed, and held at sufficient distance from the arc to prevent melting. For metals other than aluminum, the rod is held at the edge of the puddle and continuously melted off.

When a lap weld is started, the puddle is developed on the lower plate a slight distance from the joint. Immediately after the puddle is formed, move it onto the joint, holding the arc at the edge of the top sheet. Continue welding, distributing heat evenly to both plates. The puddle may sometimes be better controlled with the aid of a filler rod. With the rod being held at the correct angle, the point should be moved rapidly in and out almost flush with the bottom plate and perpendicular to the line of travel. This chilling action of the rod will provide the necessary control of the puddle.

When welding T-joints, filler metal should always be used. The puddle is formed on the base sheet away from the vertical and then moved up to the joint with the torch held at a comparatively low angle. Filler metal is added by holding the rod as nearly parallel to the vertical piece as possible and at as low an angle to the horizontal plate as is necessary to control the puddle.

With proper overlap of pieces of corner joints, the addition of filler metal is unnecessary up to ^-inch thickness.

More than one pass is usually required for welding metals of one-fourth inch or thicker.

The number of passes will be determined by the current range of the machine. The first pass must have complete fusion and proper penetration. Slightly higher amperage may be necessary when running the remaining beads of a multipass job.

In making multipass lap joints, all beads should be carried along together, staggering them so as to take advantage of heat from the previous beads. This prevents the necessity of preheating as each bead progresses.

ATOMIC HYDROGEN WELDING

Atomic hydrogen welding is a method for fusion joining of metals in which the arc is struck and maintained between two tungsten electrodes which are enveloped within a shield of hydrogen gas.

Welding heat is produced by disassociation of molecular hydrogen to form atomic hydrogen, and a recombination of the atomic hydrogen to form molecular hydrogen. This process occurs in the arc stream as current flows through the resistance of an atmosphere of hydrogen gas which envelops the tungsten electrodes.

A molecule is the smallest portion of the gas which retains chemical identity with the substance in mass. An atom is the smallest particle of a substance which can exist. To further our understanding of the definitions and processes involved in atomic hydrogen welding, let us draw a comparison between a molecule of water and a molecule of hydrogen.

As you know, there is no such thing as an atom of water— but there is a molecule of water. A combination of 2 atoms of hydrogen (H_2) and one atom of oxygen (O) is required to make the first infinitely small speck, or molecule that we can recognize as water (H_2O). Nor can hydrogen be recognized as such until 2 atoms combine to produce a molecule of the gas as we know it. Understand, however, that millions upon millions of molecules in one mass are required to make up. any volume of substance which we can use.

Assume that an empty water glass is filled with hydrogen atoms and oxygen atoms. If you get the right combination (2 atoms of hydrogen and 1 atom of oxygen) together, the glass is filled with molecules of water. If all the atoms in the glass were hydrogen, they would be "paired off" into molecules, and the glass would be filled with hydrogen gas.

It is common knowledge that hydrogen will burn. Therefore, as molecules of hydrogen gas envelope the arc flame, some of the particles burn, leaving those unburned disassociated as a gas. These unburned parts must recombine with other unburned ones to form a new molecule of the gas, which merely passes off into the atmosphere as unburned gas if they are not further consumed by the arc flame.

Heat resulting from the changing of the form of hydrogen from its molecular state to its atomic state and back again, when added to the already hot arc, creates a pinpoint concentration of tremendous heat. Hence the name, atomic

HYDROGEN WELDING.

Gas flowing from the tank, through the torch, and out through the electrode holders and tips, acts as a coolant for the electrodes. The gas also helps in transferring heat from the arc to the weld metal and provides a protective envelope about the weld area which prevents atmospheric oxidation of weld metal during the process.

EQUIPMENT

The welding machine is of the a. c. transformer type, and is designed to control and deliver electrical current from the power source to the torch, using 220- to 440-volt, 60-cycle, single-phase power supply. (See fig. 85.) It is fitted with a manual control handle for adjusting the power output, and is also equipped with an automatic hydrogen flow control on-off valve.

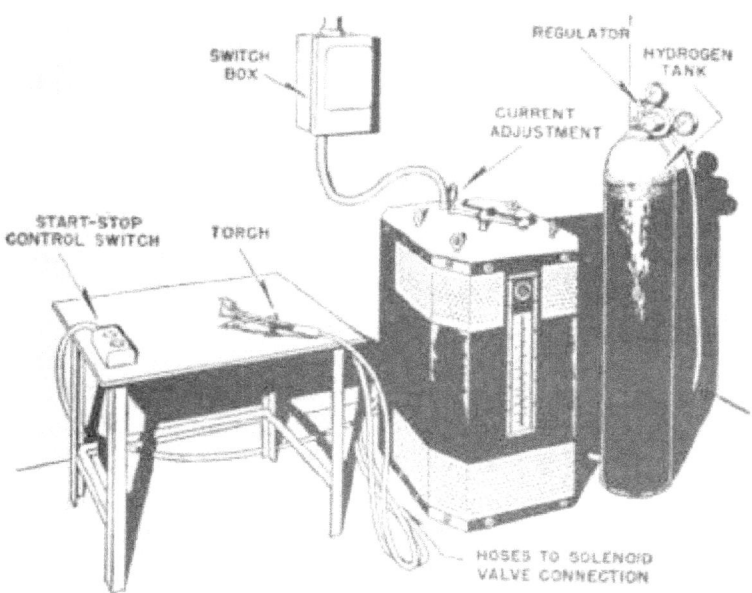

Figure 85.—Equipment for manual atomic hydrogen welding.

Alternating current is used for several reasons. Sources of power for a. c. machines are more practicable because of the ease with which available power can be converted. Electrodes are consumed more evenly, although not as much heat is generated. Power lost by the necessary methods of

controlling the output of an a. c machine is much less than that which would be lost By the d. c. machine.

Additional features of the machine, designed to improve the power factor, are: A built-in power factor correction which practically eliminates all useless lagging current; a Btepless current control to provide exact power adjustment over a wide welding range; and an auxiliary starting coil to give quick starting at low current settings. The latter is installed because an a. c arc has a tendency to start slow and be unstable at low current settings, which is one of the main disadvantages of this type of current when applied to atomic hydrogen welding.

A start and stop remote control push button functions as such, when striking and breaking the arc, to turn both the current and gas flow on and off. The control buttons, when depressed, energize or release the gas valve solenoid and the transformer switch actuators to start or stop current flow through the welding transformer.

The gas valve and its respective solenoid are a single unit which automatically turns the gas on when the start button is depressed. Some models of the machine have a gas flow adjustment on this valve. Adjustments are made by closing the valve, opening it one-half turn, then proceeding with further adjustments as necessary.

Machines are cooled by fan-forced ventilation.

HYDROGEN GAS

Hydrogen gas is an odorless, colorless, and tasteless gas which is obtainable for shop use in cylinders of 220-cubic-foot capacity charged to about 2,000 p. s. i.

Navy hydrogen bottles are painted yellow with a black band at the top.

Hand welding equipment consumes 30 to 90 cubic feet of gas per hour. Under continuous operation, a single tank will last 2y 2 to 7 hours.

Flour a So Schematic dioorom of atomic hvdraaMi wafdina ftOrth.

Electrodes

Electrode composition is identical with that used for inert arc welding, being 99.9 percent pure tungsten. Electrodes are available in diameters from 0.040 to *4 inch, and are 12 inches in length.

Electrode Holders

Electrode holders are furnished in sizes to accommodate various electrode clamps which correspond to electrode sizes. Electrode sizes to be used are, in turn, determined by the current range required for a specific job.

The electrode holder not only provides a means for establishing and maintaining the arc, but also conducts and directs hydrogen gas to shield the weld area. Current is carried by conductors through the tubes which carry gas into the electrode clamps. (See fig. 86.) Gas flow surrounding the insulated electrical conductors within the plastic handle helps to cool the entire holder, including the electrodes. One of the electrode clamps is stationary, and is cataloged as forward. The rear clamp is movable, being mounted on a coil arrangement for lateral alinement of the electrode tips.

QUIZ

1. What is inert arc welding?
2. What two gases are used in this type of welding?
3. What type of torch is used for gas shielded arc welding?
4. How do the electrodes used in inert gas welding differ from those used in other types of welding?
5. How may corrosion of electrodes be prevented?
6. How are argon cylinders identified?
7. How are helium cylinders identified?
8. What should be done when water leaks occur?
9. Should a cold electrode be allowed to touch work surfaces ?
10. In running a bead how is good penetration indicated?
11. What is atomic hydrogen welding?
12. Why is alternating current used in atomic hydrogen welding?
13. How are Navy hydrogen bottles identified ?
14. What part of an atomic hydrogen welding torch conducts and directs hydrogen to the weld area ?

APPENDIX I

GLOSSARY

Acetone —A liquid solvent used to dissolve acetylene.

Allot —A material composed of two or more elements or compounds. of which at least one is a metal. Annealing —A heat-treating operation used on metal to reduce stresses, make it less brittle, or increase its ductility. The process consists of subjecting the metal to high temperatures and then controlling the cooling.

Backstep welding (also called bark-step welding)—A method of welding in which the bead is run one section at a time in the opposite direction from the general direction of the welding.

Bead —The metal deposited during each pass of the welding process.

Ductility —The tensile property which allows a metal to be cold drawn, stamped, or hammered out thin.

Elasticity —The property which permits a metal to return to Its original shape when the force causing the change of shape Is removed.

Fatigue resistance —The ability of metallic materials to resist deterioration after shock or severe usage.

Filler bod —A small-diameter metallic rod which is melted during the welding process and added to the weld area, to supply reinforcement or to add desired qualities to the weld.

Fillet weld —A weld of approximately triangular cross-section joining two surfaces approximately at right angles to each other in a lap Joint, T-joint, or corner Joint.

Flux —Fusible material used in welding or oxygen-cutting to dissolve and facilitate removal of oxides and other undesirable substances.

Friction lighter (also called safety lighter or spark lighter)—A hand-operated device used to light the oxyacetylene torch. It consists of steel, a flint, a shield, and a spring. The spark produced burns only an instant

Hardness— The property of resisting penetration, machining, scratching, or wearing away.

Harsh flake —An oxyacetylene flame produced when the gases flow to the welding tip at a high rate of speed, fjueh a flame tends to cause the molten metal to splash around the edges of the puddle.

Jig —Any rigid structure or mechanism which holds parts while they are being welded.

Kerf —The space from which metal has been removed by a cutting process.

Mild steel —A steel with a low carbon (0.3 percent carbon and less) content.

Pickling —The process of dipping a metal in acid or other liquid to cleanse the surface of the metal.

Quenching —Cooling heated metal by sudden immersion in water, oil, or other medium.

Rosette weld —A round plug weld, used to fuse an inner reinforcing tube with the outer member.

Safety factor —The ratio of the ultimate strength of a piece of material to the actual stress or maximum permissible stress when in use.

Skip welding —Same as stagger welding.

Soft flame —An oxyacetylene flame produced when the gases flow to the welding tip at a low rate of speed.

Staggeb welding (also called skip welding)—A method of welding in which separate sections are welded, leaving gaps along the joint to be filled in later.

Stringer bead (also called string bead)—A welding bead made without side-to-side motion, as opposed to a weave bead, which is made with a weaving motion.

Tack-weld —To make a series of small welds at intervals in order to hold parts in position while they are being welded.

Tensile strength— That property of a metal which resists such forces as would tend to tear the metal apart

Tinning —Coating of solder.

Toughness —The property which allows a metal to withstand shock and to be deformed without breaking.

APPENDIX II

ARC WELDING TECHNIQUES

The following arc welding exercises furnish information regarding current, polarity, and material for performing the welding operations discussed in chapter 10, Arc Welding Processes and Techniques.

STRIKING THE ARC

LAP WELDS, FLAT POSITION

VERTICAL BEADS, WELDING DOWN

•Lap and fillet welding in vertical positions are similar. The same motions, weaves, and technique are used In both instances.

VERTICAL BUTT WELDS. WELDING DOWN

VERTICAL BUTT JOINTS, WELDING UF**«

'As this exercise concerns only beveled or V-type joints, a review of butt-joint types is

APPENDIX III

ANSWER TO QUIZZES CHAPTER 1 FUNDAMENTALS OF WELDING

1. The fusing of edges or surfaces of metals.

2. Forge welding, spot welding, resistance welding, thermit welding, arc welding, and oxy-hydrogen welding.

3. Carbon steel, chrome molybdenum, and stainless steels.

4. It is comparatively elastic, tough, and ductile.

5. Has high tensile strength, is easily machined, and is readily welded by either gas or electric methods.

6. Low heat conductivity, high coefficient of expansion, nonmagnetic properties, and satisfactory toughness.

7. The rate at which metals expand or contract This coefficient is different for different metals.

8. By tacking along the joint and placing heavy pieces of metal known as chill bars along either side of the seam.

9. The joint should be preheated to avoid uneven expansion and contraction.

10. Preheat

CHAPTER 2

INTRODUCTION TO OXYACETYLENE WELDING

1. Two cylinders, a welding torch, hose, goggles, friction lighter, wrench, and fire extinguisher.

2. Makes possible the burning of acetylene at a temperature high enough to melt the metal being welded.

3. Green or green with a white band at the top.

4. A violent explosion takes place.

5. Yellow.

6. Reduce the pressure and control the amount of gas flowing from the cylinders to the welding torch.

7. To show the pressure of the gas in the cylinder.

8. To show the pressure of the gases flowing to the torch.

9. For acetylene—50 pounds p. s. 1.; for oxygen—2,000 p. s. L 10. No. 238500'—53 18

11. The neutral flame.

12. Because it has a tendency to produce a brittle carburized surface.

13. Because it tends to form oxides causing a weak and porous mold.

14. Acetylene hose is red or maroon and oxygen hose is green or black.

15. The torch should be lighted with a friction lighter.

16. A flashback occurs when the flame disappears from the welding tip and the gases burn within the torch. A backfire occurs when the flame disappears from the tip but immediately returns. A flashback is dangerous but a backfire is rarely so.

17. In handling cylinders no valves should be opened until regulators have been installed. In moving a short distance, the cylinder may be rolled on its bottom; for longer distances suitable trucks should be used. Empty cylinders should never be used as rollers or supports.

18. Xo.

19. Oxygen should never be used as a substitute for compressed air.

20. A pressure-reducing regulator should be attached to an acetylene cylinder before acetylene is removed.

21. At all times.

CHAPTER 3

FUNDAMENTAL WELDING TECHNIQUES

1. The pull of gravity.

2. At the bottom of the weld.

3. The rod.

4. The torch is pointed in the direction of travel.

5. The torch is pointed away from the direction of travel.

6. Do not move torch ahead of puddle. Do not move flame too slowly. Do not allow molten metal to drip Into the pool.

7. Bead weld, fillet weld, groove weld, and plug weld.

8. Its limited strength and added weight.

9. The end or edge of one piece is welded to the surface of another.

10. Undercutting, poor penetration, insuftlcient reinforcement, or too much reinforcement.

11. One piece of stock lapped over the other.

12. Parts joined end to end without overlapping.

13. When load stresses are not important.

14. The face, the root, reinforcement, throat, toe, and leg.

15. 100% In a butt weld, 25 to 50% in a fillet.

16. Preheating, tacking, employing clamps, jigs or chill bars; rapid welding; backstep welding; and stagger welding.

CHAPTER 4 TECHNIQUES FOR FERROUS METALS

1. Heat-treated parts should not be welded if they cannot be reheat-treated. Gold rolled steel parts which include streamline wires, cables, tie rods or solid drawn wire should not be welded, nor should steel parts with brazed or soldered Joints.

2. Edges must be cleaned and free from all dirt, grease, oil, plating, or pa in t s.

3. It is a corrosion-resistant alloy steel which contains 18 percent chromium and 8 percent nickel.

4. Use a carburizing flame. Use a torch tip one or two sizes smaller than for plain steel. Use a flux and a filler rod of the same composition as the base metal.

5. Tacking will lessen warping and distortion.

6. Besides tacking, use clamps, copper chill plates, and jigs.

7. Use either backhand or forehand, and preheat the metal ahead of the area to be welded. Hold the rod so that the molten metal will drop into the weld. Complete the weld in one pass. Do not stop during the welding operation.

a Open single V-butt weld, open butt weld, fillet weld, horizontal fillet weld, combination sheet and the tube cluster fillet weld.

9. Preheat iron to dull red, use the proper flux, use the correct welding rod, and cool the edges of the weld very slowly.

CHAPTER 5
TECHNIQUES FOR NONFERROUS METALS

1. A metal not made from iron ore.

2. Because it is not possible to increase the strength of the weld enough to stand up to the stresses which the welded part must bear.

3. One slightly larger than that which would be used for ferrous metals of the same thickness.

4. To prevent the formation of oxide.

5. Flame adjustment, tacking methods, and rod and torch technique.

6. Flanged joint, bntt Joint, and fillet T-jolnt.

7. Equipment for reheat-treating the casting.

a When magnesium is used in a structural member.

9. No. Magnesium alloys can be welded only to other magnesium alloys.

10. Neutral. So that all oxygen will be burned and thus prevent oxidation.

11. Butt welds and fillet welds.

12. A trade name for an alloy composed principally of nickel, chromium and iron.

13. A carburizing flame.

14. Because the fumes from the melting flux are poisonous.

15. It holds the molten metal at the root of the weld.

16. Butt and fillet welds.

17. A trade name for an alloy consisting principally of nickel and copper.

18. Practically all processes.

19. Brazing.

CHAPTER 6 WELDING AIRCRAFT TUBING

1. Superior tensile and fatigue strengths, corrosion resistance, and shock impact resistance.

2. The same as for carbon steel except that the area surrounding the weld should be heated to 300-400° F. before the weld is started.

3. Low carbon (mild.steel) rods.

4. They differ in shape and in amount of strength developed.

5. A joint where a small tube is telescoped into slightly larger tube.

6. Cut with hacksaw—never with cutting torch.

7. To iucrease the shear strength of the repaired member.

8. They are sleeves that are slipped inside the structural member to provide strength without a large amount of welding.

9. An auxiliary member welded at right angles to the main member. A number of tubes welded at a common joint.

10. A gusset is a reinforcement plate between the members of a cluster or T-joint. A wrapper gusset is a double gusset made of square material. A finger strap is a patch over a damaged portion of tubing. An insert is a reinforcement that is fitted into slots in the structural members.

CHAPTER 7 CUTTING FERROUS METALS

1. It is a torch from which a stream of oxygen may be directed against metal heated by the torch flame so that the hot metal is burned away.

2. Light torch and adjust for neutral flame. Hold the torch steady. Begin cutting at the edge of the piece with torch tip vertical to the surface. Heat metal to bright red then press oxygen control lever and move torch steadily forward.

3. Cutting round bar steel, beveling steel, and cutting holes.

4. Have a fire watch stand by with a 00» fire extinguisher. Use sheet metal shields and aRbestos blankets to protect inflammable material.

CHAPTER 8 BRAZING AND SOLDERING

1. Joining metals together with bonding material of nonferrous metal or alloy whose melting point is higher than 800° P., but less than that of the metals being joined.

2. Cast iron, malleable iron, carbon steels, alloy steels, wrought iron, galvanized iron and steel, copper, brass alloys, bronze alloys, and nickel alloys.

3. Bevel the edges of the joint; clean the joint of all dirt, rust, etc., preheat the base metal; heat the rod and dip it in flux; bring filler rod near tip of torch and let the melted metal flow into the seam.

4. Use a slight excess of oxygen, clean the surface with a file or abrasive cloth, apply flux to filler rod and follow procedure as in previous answer.

5. To repair oil coolers, coolerant radlatoru, and other parts which must withstand vibration at high temperatures.

6. It must be physically and chemically clean, and the edges of the joint must fit tightly together.

7. In combination with mechanical seams; where a leakproof Joint is desired; for fitting Joints to promote rigidity and prevent corrosion; and to seal electrical connections.

8. No.

9. Heat to a bright red, clean point by filing, then dip into cleaning compound and apply solder.

10. Alloys of tin and lead.

11. Pieces must be held together while being tacked.

12. Cleaned of flux.

CHAPTER 9

ELECTRIC ARC WELDING

1. The digging quality of the arc.

2. To the current used.

3. The direction of flow of an electrical current.

4. With a voltmeter, a carbon electrode, or an E6010 electrode.

5. From the magnetic field set up around the arc, the electrode, and the work.

6. All-position, flat-position, all-position direct-current reverse polarity, all-position alternating current, all-position direct-current straight polarity, horizontal fillets and flat-positions direct or alternating current, and flat-position direct or alternating currents.

7. Flat bead, convex bead, and concave bead.

8. By a motor generator or a transformer.

9. One in which the voltage and amperage are adjustable by means of individual control.

10. Every 4 to 6 months.

11. It is an Insulated clamp which holds the electrode.

12. To protect the eyes from eye burn, and to protect the face from sunburn.

13. It protects the welder from a continuous shower of sparks and molten metal as well as the sunburn effects of the arc.

14. The steel wire brush and chipping hammer.

CHAPTER 10

ARC WELDING PROCESSES AND TECHNIQUES

1. The sticking or welding of the electrode to the work.

2. Vertical tapping motion and the scratching motion.

3. Move the electrode along the plate at a constant speed.

4. Just ahead of the crater on the work.

5. It is the movement of the electrode from side to side during the forward motion, used to get a wider deposit of welding metal.

6. Depositing more than one bead.

7. Electrode size and metal thickness.

8. Current values, electrode, polarity, arc length, and often the technique itself.

CHAPTER 11

INERT ARC AND ATOMIC HYDROGEN WELDING

1. Welding in which the arc is maintained in a shield of inactive gas.

2. Helium and argon.

3. Water-cooled torch.

4. They are made of tungsten and are practically nonconsumbale.

5. By cooling with the gas after stopping work.

6. They are painted gray with one or two white bands at the top.

7. Gray with one or two orange bands at the top.

8. Stop welding.

9. No.

10. By a very small smooth bead on the reverse side of the work.

11. Welding where an arc is maintained between two tungsten electrodes which are enveloped in a shield of hydrogen gas.

12. Sources of power are more readily available and the electrodes are consumed more evenly.

13. The bottles are painted black with a white protective cap, a white band at the top, followed by a 3-inch band of red.

14. The electrode holder.

APPENDIX IV

QUALIFICATIONS FOR ADVANCEMENT IN RATING AVIATION STRUCTURAL MECHANICS (AM)

RATING CODE NO. 810

General Service Rating

Aviation Structural Mechanics maintain and repair aircraft surfaces, structures, and hydraulic systems. Aline structural parts, such as wings, elevators, ailerons, rudders, and fuselage structures. Prepare, paint, or dope aircraft surfaces. Repair rudder, plastic, fabric, and wooden structures used in aircraft construction.

.103 BLUEPRINTS

Read simple blueprints and drawings.

Read and work from blueprints and drawings

Make working sketches for structural or hydraulic repair

.104 CONSTRUCTION

Remove, repair, service, install, and aline as appropriate aircraft structures, control rigging, and fittings, including wings, control surfaces, tabs, landing gear, control cables, and fuselage structures.

.105 CLEANING OF AIRCRAFT

Clean aircraft surfaces, structures, and enclosures, using proper materials and procedures. Use steam for cleaning aircraft

.106 METAL WORKING

Identify common aircraft metals, tubing, and fuel, oil, and hydraulic lines. Fabricate aircraft sheet metal parts, metal fittings, and tubing by cutting, flaring, bending, threading, and assembling. Use riveting tools and riveting machines. Use safety and bending wire where appropriate. Make repairs to metal structures, including stressed skin and frames

Install and maintain hydraulic lines, including the replacement of packing and seals and the repair or installation of flexible hose..

Braze, anneal, forge, and otherwise perform metal heat-treating operations encountered in aircraft structural maintenance

Use sandblasting and plating apparatus for preparing metal surfaces, if activity to which assigned is so equipped

.107 WELDING

Set up oxyacetylene welding apparatus and perform simple welding and cutting operations on carbon steel

Braze and silver-solder applicable metals. Weld ferrous and aluminum alloys

Perform simple arc-welding operations on steel plates and tubes... Note.—See welding test instructions under .400.

.108 HYDRAULIC SYSTEMS

Trace through aircraft landing gear, bomb bay, automatic pilot, brake, and other hydraulic systems; repair and service individual parts and linkages as required. Make periodic checks and inspections to facilitate preventive maintenance. Vent, bleed, drain, flush, and refill hydraulic systems. Remove, service, repair, and install hydraulic units and accessories

Set up, operate, and maintain test benches for hydraulic units and accessories

810

2, 1,C

2, 1, C

3, 2, 1, C

2, 1, C 1.C

2, 1.C

1.C

A MS

811

2, 1,C

2, 1,C

3, 2, 1, C

2, 1.C 1,C

AMH

812

3, 2, 1, C

2, 1, C

Qualifications for Advancement—Continued

Qualifications for advancement in rating

Applicable rates

AM

810

AMS

811

AMH

812

.109 WOODWORKING

Make repairs to wooden aircraft structures. Manufacture wooden forms, blocks, jigs, and templates for the manufacture or repair of aircraft structural parts.

.110 RUBBER. PLASTICS, AND FAH-RICS

Repair rubber and plastic aircraft fittings. Make repairs to self-sealing fuel cells. Perform vulcanizing operations to patch rubber material. Repair tires and tubes. Repair fabric-covered surfaces

.111 PAINTING

Prepare surfaces for painting. Mask, dope, and paint, using spray gun or brush. Mix paints according to specifications, using pigments, dryers, thinners, etc., as required. Make minor repairs to paint spray guns and accessories. Provide for proper care of equipment

.112 SAFETY PRECAUTIONS

Observe general and local safety precautions pertaining to shop and line maintenance of aircraft structures, including the precautions to be observed when painting, using power tools, welding, fueling, or otherwise servicing aircraft

2, 1,C

2, 1,C

3, 2, 1, C

3, 2, 1, C

3, 2, 1, C

3, 2, 1, C

3, 2, 1, C

3, 2, 1, C

3, 2, 1, C

Qualifications for Advancement—Continued

Qualifications for advancement in rating

.113 SUPERVISION Supervise and train personnel engaged in aircraft structural or hydraulic repair

Organize and administer:

Metal repair shop

Hydraulics repair shop

xxx .200 EXAMINATION SUBJECTS

.201 TOOLS AND MEASURING INSTRUMENTS

Types, nomenclature, and uses of hand and power-driven tools used in the structural maintenance of aircraft, including those employed in metal working, woodworking, rubber, fabrics, and plastics repairs, painting, and the rigging of cable. Types, nomenclature, and uses of various measuring instruments employed in structural maintenance of aircraft _.

.202 CONSTRUCTION

Types of aircraft construction. Maintenance procedures for removing, installing, rigging, and alining fueslage, structures, wings, tail surfaces, landing gears, tabs, control cables, cowlings, inspection plates, and fairings. Basic principles of the theory of flight and of weight and balance

Applicable rates

Qualifications for Advancement—Continued

Qualifications for advancement in rating

Applicable rates

AM

810

AMS

811

AMH

812

.203 METAL WORKING

Types, characteristics, uses, and identification markings of aircraft metals and tubing. Methods of riveting, safety wiring, and bonding. Identification of a tubing and of gasoline, oil, and hydraulic lines by AN standard markings

Processes for fabricating and joining metals. Processes and purposes of heat treating, including surface treatment for aluminum and magnesium alloys and corrosion resistant steels

.204 WELDING

Types and characteristics of welding apparatus and of welds. Welding processes, including use of material, technique, and

safety precautions

Note. —See welding test instructions under .400.

.205 HYDRAULICS

Systems of aircraft that are generally hydraulically controlled. Principles of hydraulics for transmission of power. General repair, service, and maintenance problems common to hydraulic systems, including removal, testing, installation, and inspection of various units of such systems as the landing gear, brake, flap, I

3,2, 1,C

3, 2, 1, C

3, 2, 1, C

3, 2, 1, C

3, 2, 1, C

3, 2, 1, C

3, 2, 1, C

Qualifications for Advancement—Continued

Qualifications for advancement in rating

.205 hydraulics —continued bomb bay, automatic pilot, and booster control. Principles of instruments used in aircraft hydraulic systems. Lubricants and liquids used _

.206 WOOD, RUBBER, PLASTICS, AND FABRICS

General properties of wood, rubber, fabrics, and various plastics and their uses in aircraft construction. Processes of repair, inspection, and testing of aircraft structures and fittings made from these materials

Properties and methods of preparing and applying glues and rubber. Vulcanizing processes

.207 PAINTING AND CLEANING

Types, characteristics, and properties of paints, dope, varnishes, lacquers, pigments, driers, enamels, and thinners; and effects on color and properties caused by mixing. Types of hand and air brushes used and the methods of preparing and applying paint to aircraft surfaces. Methods of cleaning and caring for painting equipment. Types, characteristics, and uses of cleaning materials for cleaning aircraft surfaces and enclosures. Materials and methods used to mask surfaces and stencil insignia or numbers on aircraft

Applicable rates

AM

810

3, 2, 1, C

3. 2, 1, C

2, 1, C

A MS

811

1,C

3, 2, 1, C

2, 1, C

3, 2, 1, C 3, 2, 1, C

AMH

812

3, 2, 1, C

3, 2, 1, C

Qualifications for Advancement—Continued

Qualifications for advancement in rating

.208 MATHEMATICS

Basic mathematical principles as that apply to aircraft structural maintenance, as follows:

Elementary principles of equations, powers, roots, and proportions. Work an elementary problem in weight

and balance

Basic principles of triangles, squares, parallelograms, circles, and functions of right angles for computing metal shapes and forms

.209 SAFETY PRECAUTIONS

Local and general safety precautions pertaining to shop and line maintenance of aircraft structures, including those to be observed when painting, doping, welding, using power-driven tools, fueling, and otherwise servicing or handling aircraft

.210 SUPPLIES

Basic principles of Navy supply system, including procurement, stowage, custody, issue, and inventory

.211 RECORDS AND REPORTS

Common forms in use and the procedures for their preparation. Records kept and reports made for administering an aircraft structural maintenance activity..

Applicable rates

238500°—53 19

Qualifications for Advancement—Continued

x x x .300 NORMAL PATH OF ADVANCEMENT TO WARRANT GRADE

Aviation structural mechanics advance to Warrant CARPENTER 7711 (Aviation Structural Technician) and assist Engineering Officers in repair and maintenance of aircraft structures.

x x x .400 INSTRUCTIONS FOR TESTING AND QUALIFYING WELDERS

Note. —Qualified welders (metal-arc and gas) are divided into three classes: welders, third class; welders, second class; and welders, first class.

.401 QUALIFICATIONS FOR WELDERS, THIRD CLASS

Pass the following qualifications test in accordance with the requirements of the General Specifications for Inspection of Material — Appendix VII — Welding, Part E: Section E-l: Test No. 1 in vertical and overhead position, using

approved electrodes. Section E-2: Test No. 1 in flat position only on steel, bronze, and

cast iron, using applicable welding rods. Section E-5: Tests Nos. 1 and 2. Pass an examination on these subjects:

Welding symbols, types of welds, nomenclature, and definitions as set forth in sections A-l and A-2 of the General Specifica-

Hons for Inspection of Material — Appendix VII — Welding, Part A.

Uses of cooper, brass, aluminum, iron, steel, and various alloys

aboard naval vessels. Preheat and postheat treatment of metals encountered in welding. Various types of metal-arc welding sets.

Current and voltage necessary for various sizes and types of

electrodes used in metal-arc welding. Proper flames and technique to be used in gas welding and cutting

of various materials, together with proper tip sizes that should

be used.

Safety precautions to be observed with regard to welding, cutting, and to handling of gases used.

.402 TESTS AND QUALIFICATIONS FOB WELDERS, SECOND CLASS

Must have served at least 1 year as welders, third class. Pass the following qualification tests in accordance with the requirements of the General Specifications for Inspection of Material — Appendix VII — Welding, Part E:

Section E-l: Test No. 4 using carbon molybdenum pipe and electrodes; Test No. 1 in flat position only, on nickel-copper, corrosion-resisting steel, and aluminum, using applicable electrodes.

Section E-2: Test No. 3 using steel tubing and welding rods; Test No. 1 in flat position on aluminum, using applicable welding rod.

Qualify to take charge of welding activities aboard ship and lay out work for men on a job.

.403 TESTS AND QUALIFICATIONS FOB WELDERS, FIBST CLASS

Must have served at least 1 year as welders, second class.

Take charge of a welding shop aboard a tender or repair ship, lay out,

and properly supervise the work. Instruct and qualify candidates for welders, third class and second

class.

.404 QUALIFICATION AND REQUALTFICATTON

The period of qualification of welders shall be for 18 months. Qualification or requalification tests will he conducted aboard repair vessels or aboard any vessel having the necessary equipment. Note : These qualifications for advancement are subject to revisions or changes at various times. These are current as of the date this book is reprinted (late 1952). However, better check to see that they are still current at the time you use this hook.

For sale by the Superintendent of Documents, U. S. Government Printing Office Washington 25, D. C. - Price $1.00